CAMBRIDGE IGCSE® GEOGRAPHY

Revision Guide

Rebecca Kitchen

ACKNOWLEDGEMENTS

Cover photo © Ruslan Kudrin / Shutterstock

Illustrations by QBS and HarperCollinsPublishers ; p.24 map sourced from Exam Extract No 717/ St Lucia © Government of Saint Lucia; p.39 top NoPainNoGain/Shutterstock; p.39 bottom Andrea Danti/Shutterstock; p.108 top right Answer section udaix/Shutterstock.

Photographs © p.14 Stacey Newman/Shutterstock.com; p.21 Adwo/Shutterstock; p.22 Graham Norris/iStockphoto; p.30 Brendan Howard / Shutterstock.com; p.32 Holger Mette/iStockphoto; p.33 Peter Treanor/Alamy; p.40 T-Tell/Shutterstock; p.42 Dutourdumonde Photography / Shutterstock. com; p.44 MarArt/Shuttestock; p.45 Karel Gallas/Shutterstock; p.52 Centre image Pablo Paul/Alamy; surrounding Caroline Green, donvictorio@o2.pl/Shutterstock, Artur Synenko/Shutterstock, ID1974/ Shutterstock, VVO/Shutterstock; p.53 501room/Shutterstock; p.60 Seaphotoart/Shutterstock; p.70 ub-photo/Shutterstock; p.73 Barry Tuck/Shutterstock; p.74 Cyrus_2000/Shutterstock.com; p.79 QiuJu Song/Shutterstock; p.82 zstock/Shutterstock; p.88 Figure photka/Shutterstock; p.90 Alf Ribeiro / Shutterstock.com; p.91 Graham Stuart/Shutterstock; p.95 arhendrix/Shutterstock

Every effort has been made to trace copyright holders and obtain their permission for the use of copyright material. The author and publisher will gladly receive information enabling them to rectify any error or omission in subsequent editions. All facts are correct at time of going to press.

Published by Letts Educational
An imprint of HarperCollins*Publishers*
The News Building
1 London Bridge Street
London
SE1 9GF

ISBN 978-0-00-821035-9

First published 2018

10 9 8 7 6 5 4 3 2 1

© HarperCollins*Publishers* Limited 2018

British Library Cataloguing in Publication Data
A CIP record for this book is available from the British Library.

Commissioned by Katherine Wilkinson and Gillian Bowman
Project managed by Sheena Shanks
Edited by Jill Laidlaw
Proofread by Louise Robb and Sonia Dawkins
Cover design by Paul Oates
Typesetting by QBS
Production by Natalia Rebow and Lyndsey Rogers
Printed in Great Britian by Martins the Printers

MIX
Paper from
responsible sources
FSC™ C007454

This book is produced from independently certified FSC paper to ensure responsible forest management.

For more information visit: www.harpercollins.co.uk/green

Contents

Theme 2: The natural environment

Theme 3: Economic development

This revision guide is designed to support your studies for the Cambridge IGCSE®, Cambridge O Level and IGCSE® (9-1) Geography. While the assessment models are slightly different for the IGCSE and O Level syllabuses this revision guide can be used for both. It contains content linked to the syllabuses, highlights the case studies that you will need to learn, provides revision tips and suggests activities and questions you might like to work through in order to prepare for the exam.

However, it is important to remember that this is a revision guide. It covers all of the syllabus content but it does not contain the depth and detail you might get from your own notes or from a textbook. It is therefore a good idea to use this book as one of the tools to help you prepare for examination but it should not necessarily be the only tool in your toolkit.

The focus of this revision guide is the content you will need for Paper 1 Geographical Themes, which is worth 45 per cent of the overall assessment. However, there is some specific guidance towards the end of the book for geographical skills and geographical investigation examinations – Papers 2 and 4 (IGCSE) and Papers 2 and 3 (O Level) (pages 93–99).

The content is arranged into the three themes that you will study – population and settlement, the natural environment and economic development. At the end of each topic area you will find quick tests and at the end of each theme you will find some exam-style questions. The quick tests are just that, quick tests to check that you have learned the main aspects of the content. The exam-style questions are reflective of the types of question you will get in the exam. They are worth 25 marks in total and to practice under timed examination conditions you should aim to complete them in about 30 minutes.

When you look at the topic areas you will notice that several of the words are in **bold**. These are key terms you will need to learn and be able to use. The definitions for these words can be found in the Glossary at the back of the book arranged in topic areas. This means that it should be fairly straightforward to find the meanings of the key words that you need.

There are revision tips that appear in each topic area. These might suggest a framework to make facts and figures easier to remember or they may highlight particular concepts that students typically get confused about. Each topic area also has a number of activities for you to have a go at. You may want to approach these in a variety of ways. For example, if you are confident with the content you might want to try some of the activities without using your notes and against the clock. However, if there are parts of the syllabus you are struggling to understand or which are difficult to learn, you could use your notes and textbook to complete the activities.

One of the important characteristics of achieving a good result is to be able to illustrate the points that you make with detailed case study examples. The case studies that you need to learn in detail are identified at the end of each chapter and an example is given.

There are many different ways in which you can revise your case studies. Firstly, it is a good idea to know which aspects of the syllabus require case studies and how the places you have studied fit with this. Then, for each case study, make sure that you have the key information to hand. You will need to learn some facts and figures for each one so make sure that this data is accurate and useful. Producing flash cards or case study factfiles are good ways of organising this information. Finally, test yourself so that you are confident that you have learned the detail of the case study. Creating quizzes, making up songs or creating diagrams from text are all good ways to demonstrate your understanding and whether or not you will be able to remember your case studies.

Finally, have a think about how you revise. While you may want to focus on your own learning, it is always a good idea to meet up with other geography students and, for at least some of the time, revise together. You can help each other with aspects one of you might not understand and devise quizzes to test each other. Most importantly, revision with someone else is usually more fun and productive than when you revise on your own.

1.1 Population dynamics

Why has the world's population increased so rapidly?

Over the last 50 years there has been a **population explosion**. The global population has risen dramatically from around 3 billion in 1950 to 7 billion in 2011. The growth happened in different places at different times but most of the most recent growth has been in poorer countries.

There are four main reasons for the population explosion, which can be remembered using the acronym **LICE**:

- Life expectancy – people are living longer due to improved medical knowledge and treatment
- Infant mortality – if there is a high infant mortality rate parents want to make sure at least some of their children survive into adulthood
- Care of the elderly – older children can look after family members
- Economic – children can be an important source of income

> **You must be able to:**
> - Describe and give reasons for the rapid increase in the world's population
> - Understand the main causes of change in population size and give reasons for contrasting rates of natural population change
> - Show an understanding of over-population and under-population and describe and evaluate population policies.

> **Revision tip**
>
> Use the acronym LICE to remember the four main reasons for the population explosion.

Activity

Sketch a graph to show world population change on the axis below. You may need to use your textbook, notes or another source in order to complete this activity.

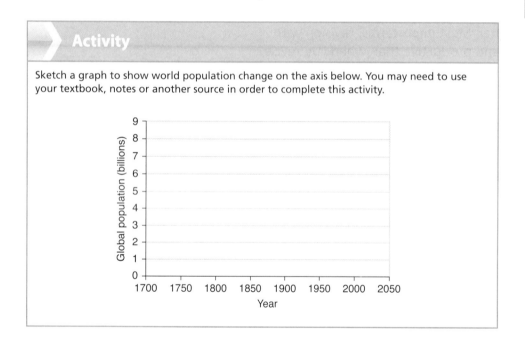

What are over- and under-population?

In some parts of the world the population is growing so rapidly that there are not enough resources, such as food and water, to go around. This is known as **over-population**.

Under-population is where there are more resources than the population needs.

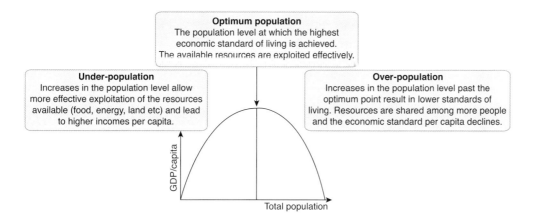

What are the main causes of change in population size?

There are three components which cause the population of an area to change in size: births, deaths and migration.

If there are more births than deaths, then the population is likely to be growing. If there are more deaths than births, then the population is likely to be decreasing. This increase or decrease is called **natural change**.

However, the picture is complicated further by **migration**. This does not add extra people to the global total but it does re-distribute them.

Activity

Complete the table below to determine whether the population of the different countries is increasing, decreasing or staying the same. The first one has been completed for you.

Country	Birth rate	Death rate	Natural change	Net migration rate	Overall change
India	19.55	7.32	12.23	−0.04	12.19
Zambia	42.13	12.67		−0.68	
Germany	8.47	11.42		1.24	
Syria	22.17	4.0		−19.79	
Russia	11.6	13.69		1.69	

Revision tip

Don't get confused between the terms over and under-populated – which are to do with availability of resources, and sparsely and densely populated – which are to do with whether there are many people living in an area or not.

If in doubt, ask yourself the following questions:

• Do people in the area have what they need in order to stay alive and to provide for their families?

• Are the country's resources sufficient to support its population?

If the answer to both of these questions is 'no' then it is likely that the area is over-populated.

Activity

'Is over-population a bad thing?' Have a discussion with a friend to try to answer this question.

Why are there contrasting rates of natural population change?

Death rates have generally decreased over the last 50 years due to improvements in healthcare. However the presence of diseases such as HIV/AIDS or malaria, natural hazards or human conflict can keep death rates high.

The very old and the very young are the most vulnerable to disease and the **infant mortality rate** is probably the most important measurement of death rates. This is because if many children die before their first birthday they will not grow to adulthood to have children of their own. A high infant mortality rate is likely to contribute to a high **birth rate** as families make sure some of their children survive.

Revision tip

There are many different factors that affect births, deaths and rates of migration and it might be difficult to remember them all. Make sure you learn at least two factors for each variable and try to explain in no more than a sentence how they cause contrasting rates of natural change.

Activity

Human conflict	Emancipation of women	Education	Urbanisation

Average age of marriage   Infant mortality rate   Good sanitation

Culture and religion   Nutrition   Clean water   Natural disasters

Sort the factors above into things that affect:
- Births
- Deaths
- Rates of migration

How successful have population policies been?

A population policy is where the government of a country imposes laws to try to control the population growth rate. **Anti-natalist policies** try to reduce births whilst **pro-natalist policies** encourage them.

An example of an anti-natalist policy is the one-child policy, which was introduced in China in 1979.

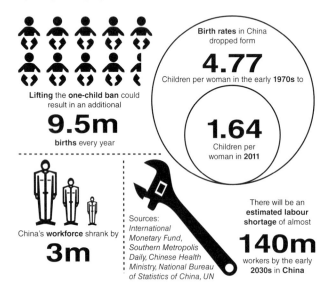

Lifting the **one-child ban** could result in an additional

9.5m
births every year

Birth rates in China dropped form

4.77
Children per woman in the early **1970s** to

1.64
Children per woman in **2011**

China's **workforce** shrank by

3m

Sources: International Monetary Fund, Southern Metropolis Daily, Chinese Health Ministry, National Bureau of Statistics of China, UN

There will be an **estimated labour shortage** of almost

140m
workers by the early **2030s** in **China**

An example of a country with pro-natalist policies is France, which has strategies such as the Carte Familles Nombreuses.

What case studies do I need?

You need case studies on:

- A country which is over-populated, e.g. Tanzania.
- A country which is under-populated, e.g. Canada.
- A country with a high rate of natural population growth, e.g. Liberia.
- A country with a low rate of population growth (or population decline), e.g. Sweden.

Quick test

1. Write down the four main causes of the population explosion.
2. Define the terms over-population and under-population.
3. Suggest **two** reasons why death rates vary from place to place.
4. Suggest **two** reasons why birth rates vary from place to place.
5. Define what is meant by a population policy and give an example.

Revision tip

You need to be able to describe and evaluate whether various population policies have been successful or not. There is not really a right or wrong answer to questions like this. Instead the examiner is looking for you to make a balanced argument to justify your decision.

Activity

- Draw a mind map to explain the benefits and problems of a population policy that you have studied.
- Write a paragraph to conclude whether you think the population policy was a success or not.

1.2 Migration

What are the different types of migration?

Migration is the movement of people. Movements can be within countries (internal migration) or between countries (international migration). **Emigrants** are people who are leaving the country whilst **immigrants** are those coming in. Some migrations can be **temporary**, the person moving may only move for a short period of time before migrating again. Other migrations are **permanent**. People may also move from rural parts of a country to the urban areas. This is called rural to urban migration.

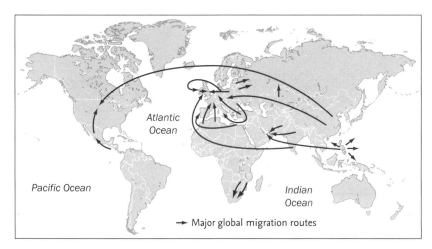

Why do people migrate?

There are many reasons why people migrate. Some of these reasons are **voluntary**, which means that the migrant has a choice whether they move or not. Many people move to find a better life for their families; they are known as **economic migrants**. However, sometimes people have no choice and are **forced** to migrate. They are known as **refugees** and often live in poverty in camps with little access to basic human needs such as food, water and healthcare.

The things that encourage people to move from the place they live in are called **push factors**, whilst the things that attract them to a different place are called **pull factors**.

> ### Revision tip
>
> Trying to classify different migrations into different types can be difficult.
>
> The most important thing is that your case study is of international migration. There needs to be clear movement between two different countries so check that your case study meets this criteria.
>
> How would you classify the following migrations?
> - From rural Luzon to Manila in the Philippines
> - From Edinburgh to North Berwick in Scotland
> - From Senegal to France

> ### Activity
>
> Fill in the missing letters in the table below to complete the examples of push and pull factors.
>
>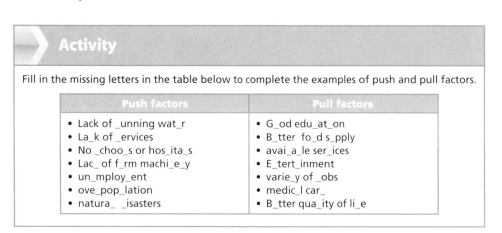

> ### Revision tip
>
> A good way to remember the difference between push and pull factors is to think about push factors pushing you out of an area and pull factors pulling you towards an area. You could act this out with your friends. Push factors are, by definition, negative and pull factors are positive. However, this doesn't mean that the place that the migrant is originating from is all bad.

What are the positive impacts of migration?

The **destination** is the place that the migrants move to whilst the **origin** is the place that they have come from. Migration can have many positive impacts on the destination country. For example:

- Migrants may be employed doing menial jobs.
- The destination country may be able to gain skilled labour cheaply.
- A multi-ethnic society may increase understanding and tolerance of other cultures.

Migration can also have many positive impacts on the country of origin. For example:

- The country can benefit from remittances, which are sent home. These can be used to improve education or healthcare.
- There is reduced pressure on resources, such as food and water, and on services such as healthcare and education.
- If the migrant returns home, they bring new skills back to the country such as the ability to speak a foreign language. These skills can help to improve the country's economy.

What are the negative impacts of migration?

Migration can also have many negative impacts on the destination country. For example:

- The children of migrants need to be educated but may not speak the language of the country.
- An increase in population may increase pressure on resources and services.
- Aspects of cultural identity may be lost, particularly if children are second generation emigrants.

Equally, migration can have many negative impacts on the country of origin. For example:

- Migrants are usually healthy young men who would be capable of doing useful work at home. A gender imbalance is created with more women than men being left behind.
- Many emigrants are educated and the population left behind are less able to build a better country.
- The young and the elderly are left behind, which puts pressure on both the education and healthcare systems.

 Revision tip

There are quite a few impacts on both the origin and destination countries which you need to learn. Some of these are social impacts whilst others are economic. You have to learn a case study of an international migration and so you may find it easier to remember these impacts if you attach them to this real life example.

Complete the table below with at least **two** impacts in each cell. Add the origin and destination of your case study.

	Origin	Destination
Positive		
Negative		

What are the positive and negative impacts on the migrants themselves?

As well as large-scale impacts on the destination and origin, migrations often have both positive and negative impacts on the migrants themselves. For example, there may be a lack of jobs available in the place they have migrated to, which may result in them facing financial problems or being exploited in low-paid jobs that they may be overqualified for. There may be pressures from home to send money back and a risk of depression if they feel isolated.

However, people often move if the pull factors outweigh the push factors and consequently they may feel positive about the migration. They may have improved job prospects, be joining family and friends or be excited by the 'bright lights' and opportunities of the place they are moving to.

Welcoming Syrian refugees to Canada

What case studies do I need?

You need a case study on:

- An international migration, e.g. Senegal to Europe.

Quick test

1. What is the difference between an immigrant and an emigrant?
2. What is a refugee?
3. Suggest **two** positive impacts that a migration may have on the destination country.
4. Suggest **two** negative impacts that a migration may have on the country of origin.
5. Briefly describe your case study of international migration. State the name of the country of origin and destination as well as **two** impacts on the migrants themselves.

Revision tip

It is sometimes difficult to learn a number of impacts on the migrant from a textbook and there are a number of ways you can bring this information to life to help you remember. One idea is to have a chat with someone who has moved house and to ask them about the positive and negative impacts that the migration has had on them. Another is to invent a migrant or take one from a story you are familiar with and try to consider the positive and negative impacts from their perspective.

Activity

Draw a picture of a migrant and annotate it to suggest how the migration may have impacted on them as an individual.

1.3 Population structure

What different types of population structure are there?

A country's **population structure** shows the percentage of people of different ages and genders. They are best illustrated using an **age/sex pyramid**. The shape of the age/sex pyramid depends on the stage of economic development that a country is at.

- Stage 1 – Remote communities: High birth rates and high death rates (population is stable).
- Stage 2 – Low income countries, e.g. Burkina Faso: High birth rates and reducing death rates (population increases rapidly).
- Stage 3 – Medium income countries, e.g. Mexico: Reducing birth rates and low death rates (population increases rapidly).
- Stage 4 – High income countries, e.g. Australia: Low birth rates and low death rates (population increases slowly).
- Stage 5 – High income countries, e.g. Sweden: Birth rates lower than death rates (population decreases slowly.)

You must be able to:
- Identify and give reasons for different types of population structure
- Describe and explain the implications of different types of population structure.

> **Revision tip**
>
> Make sure that you can sketch the shape of the age/sex pyramid at each stage of development to illustrate what the population structure is like. Don't worry about adding figures or detail but make sure that you capture the general shape. You can annotate your sketch to show your pyramid's main features. For example, high life expectancy or low birth rate.

Remote communities, Stage 1

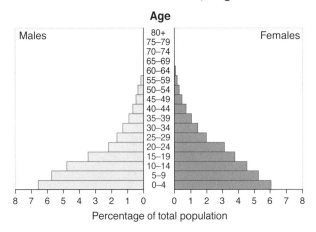

Burkina Faso, a typical Stage 2 country

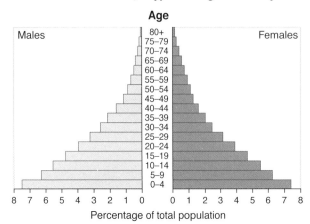

Mexico, a typical Stage 3 country

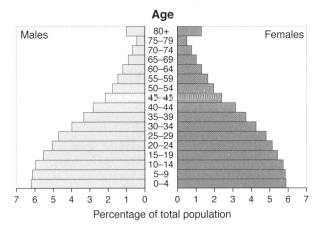

Australia, a typical Stage 4 country

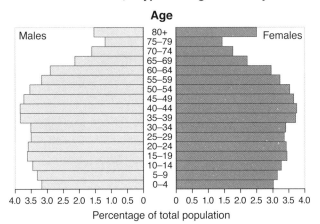

Sweden, a typical Stage 5 country

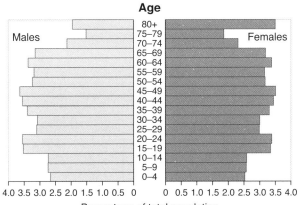

> **Activity**

Read the following descriptions and decide which age/sex pyramid is being referred to. Then make up your own descriptions for the remaining pyramids and challenge a friend to guess the correct answer.

- This pyramid does not really look like a pyramid! It shows that the population has a high life expectancy, which means that the death rate is relatively low. The birth rate is also relatively low and appears to be decreasing.
- This pyramid has steep sloping sides, which suggests that the death rate is high. The life expectancy appears to be quite low as not many people live to old age. The pyramid also has a broad base, which suggests that the birth rate is high.

What are the reasons for these different types of population structure?

Population structures depend on the birth rate, death rate and rate of migration to determine their shape. These variables are usually dependent upon the level of development of the country.

	Birth rate	Death rate
Stage 1: Remote communities	High due to: • Lack of contraception • Children work on the land • Compensation for high infant mortality.	High due to: • Disease • Poor diet and hygiene • Limited medical science.
Stage 2: Low income countries	High due to: • See stage 1.	Reducing due to: • Improvements in medical care, sanitation and water supply.
Stage 3: Medium income countries	Reducing due to: • Increased contraception • Industrialisation and mechanisation • Infant mortality falls so fewer people need to have a large family • Wealth increases and people are more **materialistic** so want fewer children.	Continues to fall due to: • See stage 2.
Stage 4: High income countries	Low due to: • See stage 3.	Low due to: • See stage 3.
Stage 5: High income countries	Very low due to: • Materialism • Desire for a small family or no children.	Low, but may rise slightly due to: • An ageing population. Deaths may increase as more people reach the end of their lives.

Revision tip

A population pyramid is descriptive, which means that, on its own, it can't explain why the population structure is as it is. However, we can suggest general reasons why it may be this shape and carry out some research into the specific country to see which of these are most likely.

Activity

For your case study of a high dependent population, explain the reasons for the shape of the age/sex pyramid. Try to add detail to your case study by giving specific dates or figures.

What are the implications of different types of population structure?

If a country has lots of young people (**youthful dependents**) or lots of elderly people (**elderly dependents**), then the country has a high dependent population. This means that there are so many people who are dependent that the working population struggles to look after them all. This is known as **age dependency**.

Youthful populations (stage 2)

- – Puts a strain on resources, such as food and water, and on services, such as healthcare and education.
- – There may be a lack of jobs available in the future.
- – A lack of education, particularly in rural areas, may lead to a large, unqualified workforce.
- + A large and cheap workforce is created.
- + A large population can provide a growing market, which is attractive to exporting countries.
- + A significant tax base may be created by the large numbers of working people.

Ageing populations (stages 4 and 5)

- – Increase in elderly people puts a strain on healthcare services.
- – Many countries face a pensions crisis where there is not enough money to cover the **pensions** of an increasingly elderly population.
- – Fewer people of working age can lead to a shrinking economy and a decrease in the amount of tax being paid.
- + Many retired people do voluntary work which is essential but carried out for free.
- + Many retired people look after their grandchildren which reduces childcare costs and allows the parents to work, which contributes to the economy.
- + Retired people still contribute a huge amount to the economy. They tend to have large amounts of leisure time and disposable income.

What case studies do I need?

You need case studies on:

- • A country with a high dependent population, e.g. Sweden.

Quick test

1. Describe what birth rates and death rates are like in a country at stage 3.
2. Sketch and annotate a population pyramid typical of stage 4.
3. Define what is meant by 'age dependency'.
4. Suggest a positive implication of a youthful population.
5. Suggest a negative implication of an ageing population.

Revision tip

It is easy to think of high dependent populations as having negative implications but there are lots of positives too. Remember to provide a balance of both in your answer unless, of course, the question specifically asks for either positives or negatives.

Activity

For each of the implications here, suggest whether it is a feature of a youthful or an ageing population and what a potential solution might be.
- • There are a large number of people requiring hip operations.
- • Unemployment is high.
- • People don't have the skills to work in service industries.
- • There isn't enough money available to provide everyone with a pension.

1.4 Population density and distribution

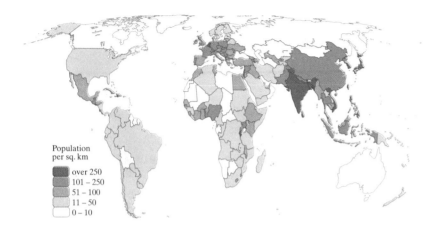

Population per sq. km

- over 250
- 101 – 250
- 51 – 100
- 11 – 50
- 0 – 10

How is the world's population distributed?

The world has an uneven **population distribution**. This means that people are spread out over the Earth's land surfaces in an uneven way. The **population density** of an area can be calculated by dividing the number of people by the size of the area (usually in km²). **Densely populated** areas are those where many people live, such as cities. **Sparsely populated** areas are those where few people live. These are often rural.

Revision tip

It is important that you can describe the population distribution of a country in a clear way. To check that you have done this, find a map that shows the population distribution of a country (you could use the map of Kenya in the exam-style questions) and write a paragraph to describe it. Make sure that you use compass directions, figures and geographical vocabulary in your description. Now, give your paragraph to a friend and ask them to sketch what they think the map looks like. They can only use your description to do this. If their sketch looks similar to the original map you have probably done a good job. If not, you may need a bit more practice.

Activity

Complete the table by calculating the population densities of the different countries.

Country	Total population	Total area (km²)	Population density
Brazil	204 259 812	8 515 770	
Egypt	88 487 396	1 001 450	
Australia	22 751 014	7 741 220	
Mali	16 955 536	1 240 192	
China	1 367 485 388	9 596 960	

Why might the population density calculation not be a particularly useful figure when looking at the population distribution of a country?

What factors influence population density and distribution?

Extreme environments are usually sparsely populated. These are areas where the climate is too hot, too cold, too wet or too dry or the terrain too mountainous for people to inhabit easily. Some areas of the world are completely **uninhabited**, usually because they are both extreme and inaccessible (difficult to get to).

Revision tip

Whilst there are many reasons why areas may be densely populated, the '5 Fs' stated here are a good starting point. You could make up a rhyme or poem to help you remember them.
• For each of the '5 Fs' explain why the presence of this characteristic might make an area densely populated.

The Valley of the Moon is part of the Atacama Desert in Chile

There are many reasons why areas may be densely populated. For example, they may have:

• flat land
• fertile soil
• fossil fuels
• fishing
• fresh water.

As well as these physical and economic factors, social and political factors affect population density. For example, people may want to live together for security or unstable political countries may have low density as people migrate.

Singapore – an example of a densely populated country

Singapore's population density is over 8100 per km². Its population density is high because it is a city state, which means that all of its population live in urban areas. A large number of people choose to live here because it is a busy trading port. It is also a successful manufacturing and financial centre. Singapore is so densely populated that many people live in high-rise blocks. The government have also built 28 new towns to try to disperse the population.

Singapore is a busy international port

Activity

Look at this photo of Singapore and write at least five clear and detailed annotations to explain why it is so densely populated.

Revision tip

Remember that just because an area is densely populated this does not mean that it is also over-populated. In fact, in the two examples shown here it is the one that is sparsely populated (the Sahel) that struggles to feed its population and where **famine** is common.

The Sahel – an example of a sparsely populated area

The Sahel is the area that lies to the south of the Sahara Desert and spans 14 countries from Senegal in the west to Ethiopia in the east. The climate of the Sahel is extreme. Temperatures are above 20°C all year round and precipitation falls in a distinct wet season which runs from May to September. The main economic activity in the Sahel is farming – millet and sorghum (food crops) and peanuts and cotton (**cash crops**). However, **desertification** and **soil degradation** are making it increasingly difficult to farm effectively.

What case studies do I need?

You need case studies on:

- A densely populated country or area (at any scale from local to regional), e.g. Singapore.
- A sparsely populated country or area (at any scale from local to regional), e.g the Sahel.

Quick test

1. What is the difference between population distribution and population density?
2. Suggest **two** reasons why an area might be sparsely populated.
3. Suggest **two** reasons why an area might be densely populated.
4. Define what is meant by desertification.
5. Define what is meant by soil degradation.

Revision tip

In many sparsely populated areas, like the Sahel, it is the natural factors such as an extreme climate that initially make it sparsely populated. However, other factors such as soil degradation and desertification, which often have a human element, often make the situation more difficult for those that live there.

SKILL – If soil degradation and desertification are important factors in your case study, make sure that you can fully describe both what they are and how they link to sparsely populated areas.

Activity

Make a case study poster to highlight the important aspects of your sparsely populated area. Include an annotated location map to show its extent and try to give several clear and detailed reasons for why it is sparsely populated.

1.5 Settlements and service provision

What patterns might settlements develop into?

There are three main patterns which **settlements** develop into over time; dispersed, linear and nucleated. **Dispersed** settlements are individual farms and houses that tend to be scattered over a wide, rural area. **Linear** settlements form a line. They are long and narrow. They tend to form along a road or river. **Nucleated** settlements occur where buildings are clustered around a road junction, church or bridge.

You must be able to:
- Explain the patterns of settlement
- Describe and explain the factors which may influence the sites, growth and functions of settlements
- Give reasons for the hierarchy of settlements and services.

> ### Activity

Look at the map extract.

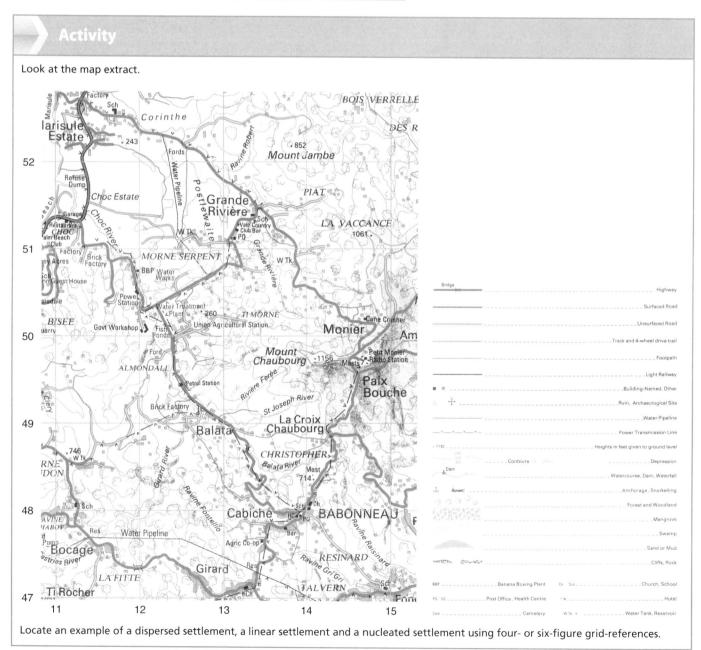

Locate an example of a dispersed settlement, a linear settlement and a nucleated settlement using four- or six-figure grid-references.

Revision tip

Locating the different types of settlement on a map is a useful way to show your understanding. Make sure that you know how to work out both four- and six-figure grid references so that you can give your locations clearly. It makes most sense to use six-figure grid references for dispersed settlements and four-figure grid references for linear and nucleated settlements as these tend to take up more space.

What factors influence the site, growth and function of a settlement?

The **site** of a settlement is the land on which a settlement is built. A good settlement site is one which has a number of useful physical factors. If a settlement is also accessible and has a good **situation**, then it is likely to grow in size and have more varied **functions** (activities).

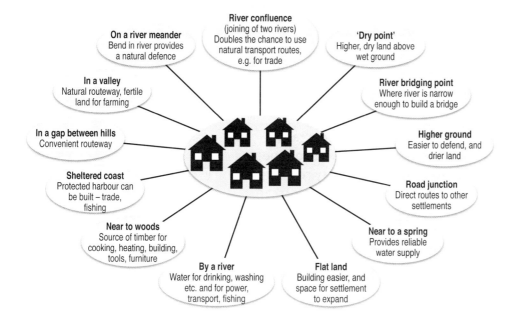

Revision tip

Over time the importance of site factors has changed. Hundreds of years ago, having a site on a hill or on the inside of a meander would be beneficial as it would be easily defended from attack. Also, having access to a water supply would be important. In many countries today the risk of attack is reduced and water is ubiquitous (it's everywhere) meaning that these site factors are perhaps less important than they used to be.

Activity

Design a poster that will help you to remember how specific physical factors (such as relief, soil and water supply) influence the site of a settlement.

Why do settlements and services have a hierarchy?

We can order settlements according to their functions and population size. This is known as a **settlement hierarchy**.

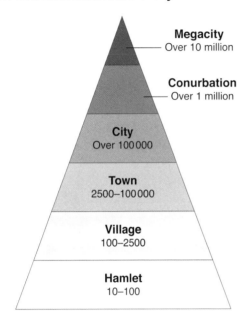

Megacity
Over 10 million

Conurbation
Over 1 million

City
Over 100 000

Town
2500–100 000

Village
100–2500

Hamlet
10–100

Generally, the larger the population, the more services a settlement provides and the higher up the hierarchy it is. For example, villages near the bottom of the hierarchy may be able to support a primary school, small general store and a bus stop. However, large cities, at the top can support universities, cathedrals and airports.

The number of people needed to support a service is its **threshold population**. For example, a primary school needs a threshold population of 500 people whereas a university needs a threshold population of 100 000 people. Because each type of service has its own threshold population, it requires a minimum area from which enough people can be drawn. This is known as its **catchment area** or **sphere of influence**. Low order goods (cheap and bought frequently) such as bread and newspapers have a small sphere of influence. High order goods (expensive and bought rarely) such as beds and cars have a large sphere of influence.

SKILL – One way to investigate the threshold population of different services is to carry out a land use survey. You could visit a number of settlements in an area and classify the different services that you spot. Examples of classification headings could be: food, medical, financial, educational, shops (excluding food), other. How are the different services distributed? How does this link to where the settlements appear in the hierarchy?

Revision tip

It is easier to remember what a settlement hierarchy looks like if you think about settlements that you know well and consider where they would appear in the hierarchy. For example, you might think about what services your local village, town or city has which you use.

Activity

Sketch a settlement hierarchy and suggest an example for each of its layers.

Revision tip

Threshold populations are a bit more complicated than the information here suggests. For example, not all people have the same needs. In a place with a large number of retired people a shop selling aids for the elderly would be much more likely to succeed than, for example, a toy shop.

What case studies do I need?

You need case studies on:

- Settlement and service provision in an area, e.g. the Isle of Wight.

Quick test

1. Sketch a simple diagram to show a linear settlement.
2. What is the difference between settlement site and situation?
3. Define what is meant by a threshold population.
4. What is a likely threshold population for a primary school?
5. What settlements in the hierarchy would you expect to be able to support a primary school?

Activity

Write a shopping list that includes both high order and low order goods. Now look at the settlement hierarchy you sketched previously. Where in the hierarchy would you go to buy each item on your list? Why? Are there any answers that surprise you?

1.6 Urban settlements

What are the characteristics of different land use zones found in urban areas?

Parts of urban areas can be classified into different **land use zones**. These describe the main function of the area and the way in which the land is used. There are four main types of urban land use zones in MEDC cities.

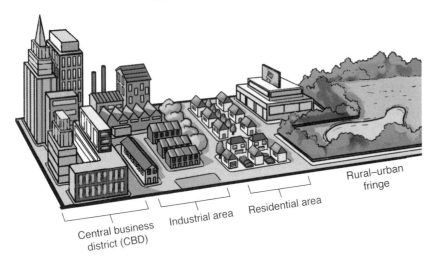

Central business district (CBD) Industrial area Residential area Rural–urban fringe

You must be able to:
• Describe and give reasons for the characteristics of, and changes in, land use in urban areas
• Explain the problems of urban areas, their causes and possible solutions.

- **The Central Business District** (CBD) – The economic centre of an urban area where the road and railway networks meet. This zone has good access. It is a good location for major public buildings such as cathedrals, theatres, banks, offices and shops selling comparison goods. The CBD is the oldest part of the urban area and the area where land is most expensive.
- Industrial areas – In more economically developed countries industry is usually located towards the outskirts of the urban area. Light manufacturing and retail industries often cluster together on land that is relatively cheap and has good transport links.
  In less economically developed countries the industrial areas tend to be located along main roads so that they are accessible.
- Residential areas – In more economically developed countries the land use towards the centre of the urban area (inner zone) dates from the Industrial Revolution. Houses were back-to-back terraces and located next to the factories where people worked. However, much **redevelopment** has taken place in this area. Low quality housing and old, heavy industry has been removed. Ring roads, retail parks, cinemas and car parks have taken their place. Towards the edge of the urban area the land is cheaper, which means people can afford large plots with gardens and garages. Many people living in the suburbs have a high **quality of life** and **commute** to work.
  In less economically developed countries high quality housing is usually found near the centre of the urban area. The people that live here want to be close to the attractions of the CBD. The outskirts of the urban area are dominated by **squatter settlements**. These are characterised by low quality housing and poor **infrastructure**.

Revision tip

You need to make sure that you know the characteristics of each zone. You will also need to be able to recognise each zone from photographs, maps and street plans.

Activity

Find four photographs; one of each land use zone. Annotate each photograph to describe the characteristics which are found in each zone.

- **The rural–urban fringe** – This zone is where the urban area and the countryside meet. A mixture of different land uses is common. For example, you would expect to find motorways, landfill sites, golf courses, out-of-town shopping, business parks and some housing and farms in this zone.

Why does the land use of urban areas change?

The proportion of the world's population living in urban areas has grown over the last 200 years. This process is called **urbanisation**. Whilst urbanisation originally occurred in more economically developed countries, today most urbanisation is happening in less economically developed areas of the world. More people living in urban areas has led to **urban sprawl**. Urban sprawl is where the urban area starts to grow into the surrounding rural area, changing the land use from agricultural to residential or industrial.

What are some of the problems found in urban areas?

Rapid urbanisation has led to a range of problems in urban areas.

- Inequalities – Inequalities are differences in wealth and quality of life. In more economically developed countries people living near the centre of the urban area often have a poorer quality of life than those who live in the outskirts.
- Traffic congestion – As more people live on the edge of urban areas and commute to work, so traffic congestion increases.
- Housing issues – Building new and affordable homes in urban areas is difficult. This is because the price of land is high and in short supply.
- Conflicts over land use – The mixed land use of the rural–urban fringe places different demands on the area and often leads to conflict. For example, the quarrying of stone can cause pollution and congestion that affects people living nearby. Urban sprawl can also put pressure on the Green Belt (an area of land which is protected and has restrictions on development).
- **Pollution** – There are four main types of pollution: air, noise, water and visual. All are present in urban areas but air pollution is perhaps the most serious. **Smog** is particularly dangerous to human health and can cause breathing problems.

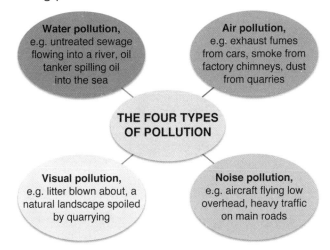

Revision tip

It's a good idea to link your ideas in this section with those in section 1.1, which explains why the population of the world has increased dramatically, and section 1.2, which mentions rural to urban migration.

Activity

Explain why urbanisation is occurring, particularly in less economically developed areas of the world.

Revision tip

Whilst you will find all of these problems throughout an urban area, they tend to be concentrated in particular parts. For example, traffic congestion is mainly a problem in the CBD and housing issues are a feature of the outskirts of urban areas. This information will add detail to your answer and also mean that you can suggest more specific solutions to these problems.

What might some of the possible solutions be?

- Inequalities – Redevelopment can improve the physical environment and the quality of housing in a run-down area. However, sometimes this can create even greater inequality as local residents can no longer afford to live in the area.
- Traffic congestion – There are a number of strategies that can ease traffic congestion. Park and Ride encourages people to park their cars on the outskirts of an urban area and then take a bus into the centre, whilst **congestion charges** can discourage people from entering the busiest parts of an urban area.
- Housing issues – There are two types of site on which houses can be built; **greenfield** sites and **brownfield** sites. Building on brownfield sites reduces urban sprawl and can improve the urban environment by building on derelict sites. However, building on these sites is likely to be more expensive than building on greenfield sites.
- Conflicts over land use – Increased information or consultation can reduce conflicts. For example, signs informing visitors may make them less likely to damage farmers' crops.
- Pollution – Creating pedestrianised zones can reduce pollution in urban centres by discouraging traffic.

Activity

Draw a picture to help you remember the different problems that are experienced in urban areas. Add to your drawing or add annotations to suggest how these problems could be solved.

Revision tip

Make sure that you can think of an example where each of the solutions has been implemented. For example, the Metro Monorail in Sydney, Australia, is a good example of a **rapid transit system** implemented to reduce traffic congestion.

A pedestrian zone in Edinburgh, Scotland, where traffic access is limited

You need case studies on:

- An urban area or areas, e.g London. Particularly the characteristics of, and changes in, the land-use, problems of the urban area and the solutions that have been adopted.

Quick test

1. Name the **four** main types of urban land use zone.
2. Describe **two** characteristics of residential areas found on the outskirts of urban areas.
3. Define what is meant by pollution.
4. Suggest **two** other problems found in urban areas.
5. For each problem you have identified, suggest an appropriate solution.

1.7 Urbanisation

What are the reasons for rapid urban growth?

The process of **urbanisation** is occurring in both more and less economically developed countries. However, urban areas are growing fastest in less economically developed countries. This is mainly a result of rural depopulation as people engage in rural to urban migration. **Over-population** and a reduction in crop yields in rural areas are **push factors**. A higher standard of living and access to better paid jobs in urban areas are **pull factors**. Migrants also tend to be young and have higher birth rates and life expectancies, which further increases the population.

What are the impacts of urban growth on rural and urban areas?

As urbanisation occurs, so urban sprawl increases and encroaches into the rural–urban fringe. Rural landuse becomes increasingly urbanised, which changes its character and nature. These impacts are more fully described in section 1.6. In less economically developed countries urbanisation as a result of rural to urban migration has led to the increase of squatter settlements.

Shacks in squatter settlements are made of any available materials

You must be able to:
• Identify and suggest reasons for rapid urban growth
• Describe the impacts of urban growth on both rural and urban areas, along with possible solutions to reduce the negative impacts.

Revision tip

Many of the words and concepts covered in this section are mentioned in previous sections. Make sure you make the links to join up your knowledge. It will make your revision more efficient and enhance your understanding.

Activity

Draw a cartoon strip to explain the reasons for rapid urban growth.

Some of the characteristics of squatter settlements include:

- Poor building construction – squatter settlements are illegal. Buildings are constructed from basic materials such as scrap corrugated iron and discarded wooden pallets.
- Disease – there is limited sewage and waste disposal. Water quality is poor and disease is rife.
- Risk of fire – houses are built close together and from flammable materials.
- Unsuitable location – squatter settlements are located on poor quality, marginal land.
- Crime – due to the poor quality of life, stress levels are high. Many people are unemployed and turn to crime in order to survive.
- Lack of healthcare and education – as the settlements are illegal the city authorities do not provide healthcare and education.
- **Informal labour** – due to the lack of education in squatter settlements, many migrants are illiterate. Young children are sent onto the streets to earn money by shining shoes and peddling cheap goods.

What are some solutions to reduce the negative impacts?

Examples of solutions to some of the problems are outlined in section 1.6. However, in relation to squatter settlements, there are two main ways that families can attempt to improve their quality of life.

- **Self-help schemes** – city authorities support families to improve their homes. They provide grants, building materials, loans and water standpipes to share. Families often buy the land to make their investment more worthwhile. Communities also may be assisted in building schools and health centres. Self-help schemes have been effective in Roçinha, a favela in Rio de Janeiro, Brazil.

A self-help scheme is Roçinha, Rio de Janeiro

- **Site and service schemes** – these are on a larger scale than self-help schemes. Water, sanitation and electricity are provided to each plot before the start of building. People then use the materials they can afford at the time. Improvements to buildings can be made later. An example of a site and service scheme is 10th of Ramadan City, near Cairo, Egypt.

Revision tip

The conditions in squatter settlements are challenging but it's important to remember that people living there do have access to basic services. Don't write that people have 'no education', 'no healthcare' or 'no sanitation' as this is unlikely to be true.

Activity

Write down as many impacts of urban growth as you can. Firstly, colour code them to show positive impacts and negative impacts. Then sort the impacts under three headings – physical, economic and social. You may find that some impacts fall under more than one heading. Finally, decide which impacts refer to the urban area, the rural area or to both.

Curitiba in Brazil is a good example of a city which has attempted to solve the problems of rapid urbanisation. It has put in place several **sustainable** or 'green' strategies.

 Revision tip

Make sure that your case study notes cover not just the problems and characteristics of squatter settlements but also how they have been improved. It is important to be clear who has been responsible for this improvement. Is it the local government, the people living in the settlement or a combination of the two?

| 1960s – Curitiba's 'Master Plan' approved. Its aim was to control urban sprawl, reduce traffic congestion and provide affordable public transport. | 1980s – 'Green zones' created to protect from unsustainable developments, 1.5 million trees planted in deprived areas, 17 urban parks established. | 2000s – New technology park to research non-fossil fuels, highest recycling rate in the world, 150km of urban pathways for walkers and cyclists. |

| 1970s – Shopping streets pedestrianised, bus-only lanes introduced, streets made one-way, industrial areas established on outskirts of city. | 1990s – Botanical gardens created, buses capable of carrying 270 passengers introduced, the Bus Mass Transit system reduces car journeys by 70%. |

 Activity

You could construct a model of a squatter settlement in a shoe box using cardboard, string and other bits and pieces. One half of the model could show the characteristics of the squatter settlement and the other could show the improvements that have been made to it. Don't forget to annotate your model to describe the characteristics and improvements in detail. It's always a good idea to base your model on a real example and to make it as factually accurate as possible.

What case studies do I need?

You need case studies on:

- A rapidly growing urban area in a developing country and migration to it, e.g. Mumbai.

Quick test

1. Why has rapid urbanisation occurred in less economically developed countries?
2. Give **two** characteristics of squatter settlements.
3. Give an example of a squatter settlement.
4. What is the main difference between self-help schemes and site and service schemes?
5. Suggest **one** solution to the negative impacts of urbanisation.

 Revision tip

Whilst it's quite useful to remember a couple of dates so that you have an idea of the time frame, it is more important to remember the strategies. Make sure that you learn several of these and that you can explain why they are sustainable.

 Activity

Write a paragraph to explain why Curitiba is a good example of a sustainable city. Do you think that other cities could put in place similar strategies? Explain your answer.

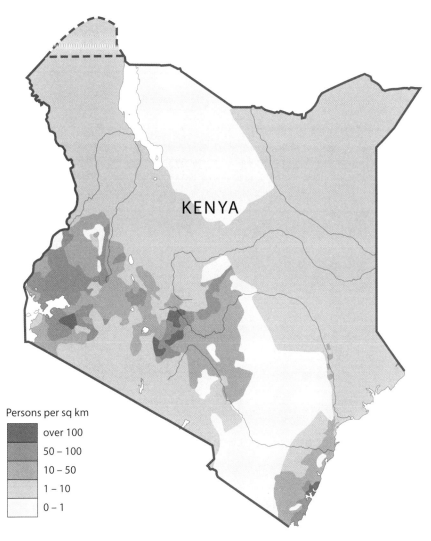

Persons per sq km
- over 100
- 50 – 100
- 10 – 50
- 1 – 10
- 0 – 1

1 This map shows population density in Kenya. Which part of Kenya has the lowest population density?

.... **[1]**

2 Migration affects many people globally both directly and indirectly. Define what is meant by 'migration' and give an example of a migration.

....

.... **[2]**

3 Describe the patterns that settlements typically develop into.

....

....

.... **[3]**

4 Look at the data about Kenya in the table below.

Birth rate	Death rate	Infant mortality	Net migration rate	Life expectancy
26.4	6.89	39.38	−0.22	63.77

Calculate the population growth in Kenya in 2015.

[3]

5 Sketch and annotate a population pyramid which shows the population structure for a country at Stage 4.

[4]

6   Urban areas (such as Beijing) are home to over 55% of the world's population. Describe some of the problems found in urban areas and suggest some possible solutions.

....

....

....

....

....

....

....

....

....

....

....   [5]

7   Choose an urban area you have studied which is rapidly growing. Explain the characteristics of this urban area and some of the things that have been done in order to make it more sustainable.

....

....

....

....

....

....

....

....

....

....

....

....

....

....

....   [7]

What are the main types and features of volcanoes and earthquakes?

Volcanoes can be classified in terms of their activity. Active volcanoes erupt frequently. Dormant volcanoes are not currently erupting but could do so in the future. Extinct volcanoes are unlikely to erupt again.

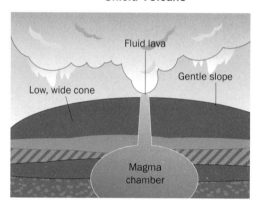

Shield Volcano

Volcanoes can also be classified in terms of their shape. The two main types are **shield volcanoes**, which are relatively flat and have gently sloping sides. The **lava** that forms a shield volcano is thin and runny and comes from frequent and gentle eruptions. **Strato volcanoes** are made up of alternating layers of ash and lava. The lava that forms strato volcanoes is thick and sticky and comes from infrequent but violent eruptions. **Pyroclastic flows** are also common.

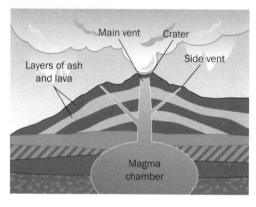

Strato volcano

Volcanoes have a number of features. These include a **magma chamber**, a **main vent**, side vents and a **crater**.

You must be able to:
• Describe the main types and features of volcanoes and earthquakes
• Describe and explain the distribution of earthquakes and volcanoes
• Describe the causes of earthquakes and volcanic eruptions and their effects on people and the environment
• Explain what can be done to reduce the impacts of earthquakes and volcanoes.

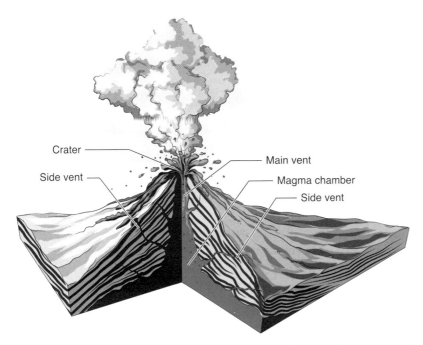

Crater — ⎯ Main vent
Side vent — ⎯ Magma chamber
⎯ Side vent

Revision tip

It can be easy to get mixed up with the different types of volcano and features of earthquakes. One method that may help you to remember is to draw the diagrams showing the features from memory, which you can then label. Another way is to create a poem or a song, which will help you to remember all of the vocabulary.

Earthquakes also have a number of features. The **focus** is the centre of the earthquake, deep under the Earth's surface. **Seismic waves** move outwards from the focus, where they are most intense and cause tremors in the ground. The **epicentre** is the point on the Earth's surface directly above the focus.

Activity

Use modelling clay or building blocks to create a model of a shield volcano. Make sure you can point out all of its features. Then create a model of a strato volcano. What are the main similarities and differences between the two models?

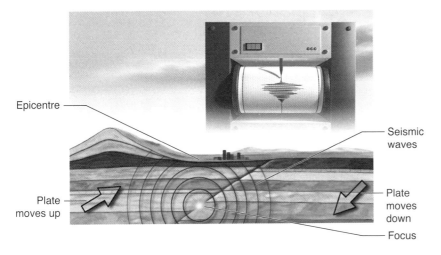

Epicentre —
Seismic waves
Plate moves up
Plate moves down
Focus

How are earthquakes and volcanoes distributed?

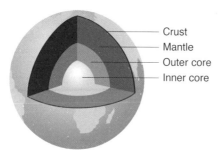

Crust
Mantle
Outer core
Inner core

The Earth is made up of a series of layers; the **inner core** and the **outer core**, the **mantle** and the **crust**. There are two different types of crust – oceanic and continental – and it is broken up into pieces called **tectonic plates**. These plates float on top of the mantle and move as a result of **convection currents** within it. Consequently, the plates move very slowly; moving together, pulling apart or rubbing against each other.

The majority of volcanoes and earthquakes are found along **plate boundaries**.

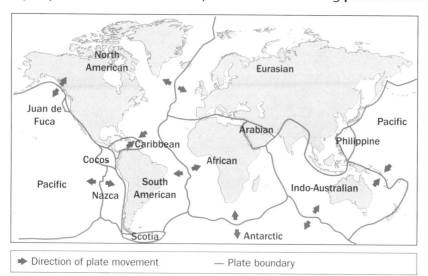

→ Direction of plate movement   — Plate boundary

> ### Revision tip
>
> You won't be expected to know or to remember all of the Earth's tectonic plates. However, it is quite useful to have a couple of examples up your sleeve. Try to pick some that you will remember easily such as the plate on which you live or one that crops up in a case study.

> ### Activity
>
> On a world map showing the plates (such as the map here) highlight the plate boundaries that your case studies can be found on (have a look at 'What case studies do I need?' if you need a reminder). Add arrows to show the direction of movement of these plates. What types of plate boundary are they?

A rift in the plate boundary between the North American and Eurasian plates in the Thingvellir National Park, Iceland

What are the different types of plate boundary?

Where the tectonic plates are moving together this is known as a **destructive** plate boundary. Here, continental crust meets oceanic crust. The oceanic crust is more dense and is forced under the continental crust in an area called the **subduction zone**. Friction occurs as the plate moves downwards, which causes an earthquake. The increase in temperatures also causes the crust to melt. The resulting **magma** is forced to the surface to form a volcano. A good example is the Andes of South America.

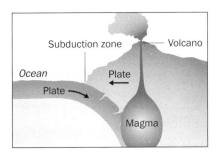

Where the tectonic plates are pulling apart this is known as a **constructive** plate boundary. Here, a gap is created between the two pieces of crust and magma rises to fill this gap. This creates new oceanic crust. A good example is the mid-Atlantic ridge.

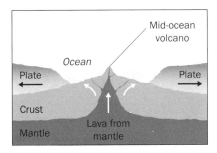

Plates can also move in the same direction but at different speeds.

Where the tectonic plates are rubbing against each other, this is known as a **conservative** plate boundary. Here, there is no subduction and therefore no volcanoes are formed. However, if the plates become stuck, pressure can build. If this pressure is suddenly released it will cause an earthquake. A good example is the San Andreas Fault in California, USA.

How can earthquakes and volcanoes affect people?

Living near a volcano can bring many benefits. Ash from the volcano creates fertile soil that is excellent for growing crops. Also, volcanoes create opportunities for tourism, which can generate income and employment. Heat from the earth can be used for geothermal energy production and minerals are often a feature of volcanic areas, which can be mined.

However, both earthquakes and volcanoes are also hazardous. When an earthquake occurs the shaking of the ground can cause buildings to collapse. This can kill and injure people who become trapped. Earthquakes are very difficult to predict and so it is hard to evacuate people. Consequently, some buildings are built to be earthquake proof. This means that they are less likely to collapse in an earthquake.

Volcanoes are easier to predict and tend to be closely monitored. This means that people can be evacuated before an eruption. This can save lives, although buildings and infrastructure may still be destroyed.

Often the death toll from earthquakes and volcanoes is greater in less economically developed countries. This is because the buildings in these countries are likely to be poorly built and healthcare services may be scarce.

> **Revision tip**

Earthquakes and, to a lesser extent, volcanoes regularly make the news. Have a look at a recent event of this type to see how people were affected – this could become your case study for this unit. Where did the event occur? How were people affected? Why do you think they were affected in this way?

A house damaged by the earthquake in Nepal in April 2015

What case studies do I need?

You need case studies on:

- A volcano, e.g. Montserrat.
- An earthquake, e.g. the Tohoku earthquake in Japan in 2011.

Quick test

1. Sketch a strato volcano and label the main parts.
2. Give **two** characteristics of the Earth's crust.
3. What is a constructive plate boundary?
4. Give **one** advantage of living near a volcano.
5. Suggest **one** economic impact of an earthquake.

Activity

Create a fact file for your earthquake case study. Include a map showing the epicentre of the earthquake and a number of ways in which the earthquake affected people. Try to classify these into social, environmental or economic impacts.

2.2 Rivers

What are the main characteristics of a drainage basin?

A **drainage basin** is the area drained by a river. All of the water in the drainage basin eventually ends up in the river. There are many processes which occur during this journey from drainage basin to river:

- **Interception**
- **Infiltration**
- **Overland flow**
- **Evaporation**
- **Throughflow**
- **Groundwater flow**

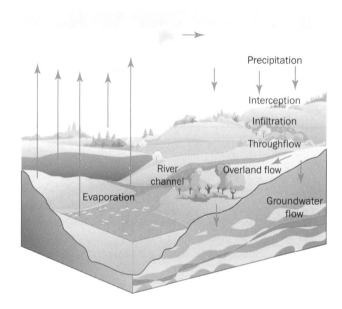

The drainage basin has a number of characteristics. The watershed marks the boundary of the drainage basin and is the highest land. The source, or beginning, of the river is found in these upland areas. Near the source the river is narrow and shallow. The gradient of the land is steep and so the river also has lots of energy. Many of these small rivers are tributaries or branches of the main river. A confluence is the point where tributaries join. As the river flows towards its mouth and more tributaries join the main river, so it becomes wider and deeper. The river's discharge – the amount of water flowing – also increases towards the mouth.

How do rivers erode, transport and deposit?

Where the river has high energy (near the source) erosion occurs. The river then transports this eroded material, its **load**, until it is deposited towards the mouth. There are four main processes of river erosion: **hydraulic action, corrasion, attrition** and **corrosion.**

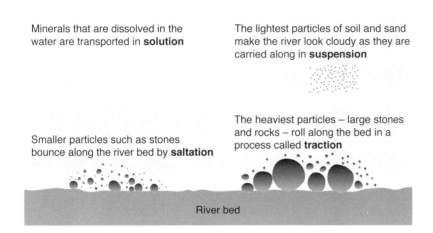

Minerals that are dissolved in the water are transported in **solution**

The lightest particles of soil and sand make the river look cloudy as they are carried along in **suspension**

Smaller particles such as stones bounce along the river bed by **saltation**

The heaviest particles – large stones and rocks – roll along the bed in a process called **traction**

River bed

The river can transport this eroded material in a number of ways. Tiny particles of sand and silt are dissolved or flow in the water as suspended sediment. Larger pebbles and boulders will bounce or be rolled along the river bed.

As the river loses energy it starts to deposit its load. The larger pebbles and boulders are deposited first as they are heavier and need more energy to move them. Most deposition occurs towards the mouth of the river. However, along the long profile there will be places where the river slows down and deposition occurs.

What landforms are formed by the process of erosion, transportation and deposition?

Near the source, the river occupies a V-shaped valley. The load is large as it has not had time to be eroded. There is friction between the water and the load, which can slow the river's velocity. The main landforms of erosion, which are found near the source, are waterfalls and potholes.

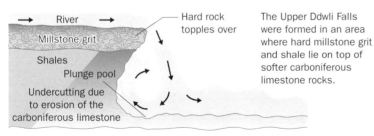

River

Millstone grit

Shales

Plunge pool

Undercutting due to erosion of the carboniferous limestone

Hard rock topples over

The Upper Ddwli Falls were formed in an area where hard millstone grit and shale lie on top of softer carboniferous limestone rocks.

The formation of the Upper Ddwli Falls, South Wales

In the middle course of the river the valley sides are more gentle. Lateral (side to side) erosion is common and the river starts to meander. The main landforms are meanders and ox-bow lakes.

Potholes forming in a river

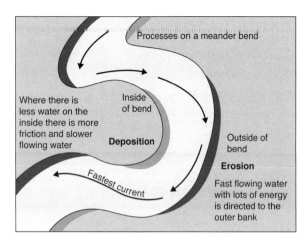

Processes on a meander bend

Inside of bend

Where there is less water on the inside there is more friction and slower flowing water

Deposition

Outside of bend

Erosion

Fast flowing water with lots of energy is directed to the outer bank

Fastest current

It is much easier to describe the formation of river landforms using a diagram than trying to describe it with text. Make sure that you know how to draw a diagram or series of diagrams to show the formation of each of the landforms.

Activity

Play taboo with all of the vocabulary you have had to remember so far in the river topic. Write each key word in bold on a card and then write three other words associated with the key word below. The idea is that you have to describe the key word without using either it or any of the three associated words.

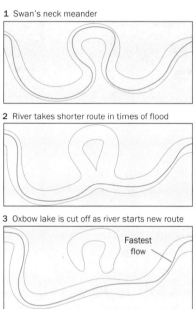

1 Swan's neck meander

2 River takes shorter route in times of flood

3 Oxbow lake is cut off as river starts new route

Fastest flow

Deposition occurs at the ends of the oxbow lake

Near the mouth of the river, the river is fast flowing because there is little friction with the bed, banks and load. However, the river is losing energy and so deposition occurs. The river continues to meander over its flood plain, which is the area that floods if the river bursts its banks. The channel's edges have natural ridges called levees which get higher with each successive flood. As the river enters the sea it may form a **delta**. It may also end in an estuary, which is wide and deep.

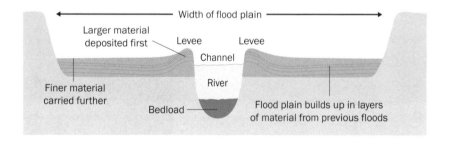

Width of flood plain

Larger material deposited first

Levee

Levee

Channel

River

Finer material carried further

Bedload

Flood plain builds up in layers of material from previous floods

What opportunities and hazards do rivers present?

Rivers present a number of opportunities. Erosion near the source of the river can create spectacular landforms such as waterfalls which can become tourist attractions. Towards the mouth, the soils on flood plains and deltas are fertile and excellent for growing crops. Also, industries which require large sources of water, such as hydro-electric power plants, need a river location and particularly, a steep gradient. Rivers are also transport routes for goods and provide recreational opportunities such as fishing.

However, rivers are also hazardous and may become more so as a result of climate change. Rivers are liable to flooding and people who live on flood plains are particularly at risk. River flooding tends to be caused by heavy rain. Water reaches the river channel more quickly, causing it to burst its banks and flood. In order to manage river flooding, hard or soft engineering can be undertaken. Hard engineering projects involve construction, tend to be expensive and have a greater impact on the river and its landscape. Examples of hard engineering include the construction of dams to control discharge or engineering to widen and deepen the river channel. Soft engineering is usually more natural and sensitive to the landscape. Examples of soft engineering include afforestation, where trees are planted to intercept rainfall, and managed flooding where the river is allowed to flood in some places to prevent flooding in others.

What case studies do I need?

You need case studies on:

- The opportunities presented by a river or rivers, the associated hazards and their management, e.g. either the Mekong delta, Vietnam, or the Mississippi River, USA.

Quick test

1. Define the term 'overland flow'.
2. Name **two** different types of river erosion.
3. Give an example of a landform you would expect to find in the middle course of a river.
4. True or false? A levee is a man-made barrier that prevents flooding.
5. Suggest **one** opportunity and **one** hazard of rivers.

Revision tip

Make sure you learn the detail for your case study which illustrates this section. Write all of the facts and figures that make up the case study on bits of paper; then fold them up. Pick a piece of paper at random and try to remember what the fact or figure relates to. Can you remember more than five facts or figures?

Activity

Sketch a map of the river in your case study. Shade and annotate the map to show the locations of various opportunities or the area that is most likely to be affected by the hazards.

2.3 Coasts

How do waves erode, transport and deposit and what landforms are formed by these processes?

There are two different types of waves. **Constructive waves** are low and gentle. They have a strong **swash** and weak **backwash** and build up beaches. **Destructive waves** are high and powerful and are the result of windy conditions and a large **fetch**. They have a strong backwash and weak swash and remove material from beaches.

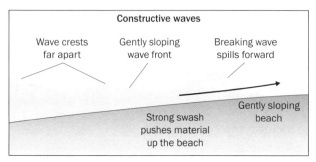

Swash is stronger than backwash, so waves run gently up the beach – material is carried onto the beach and deposited there.

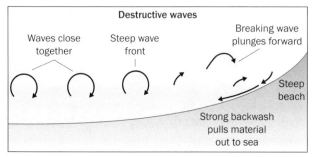

Backwash is stronger than swash, so waves crash onto the beach – material is eroded from the coastline.

Coastal erosion involves the same processes as river erosion: hydraulic action, corrasion, corrosion and attrition. Weathering processes also break up rocks. The eroded material is transported by a process known as **longshore drift**.

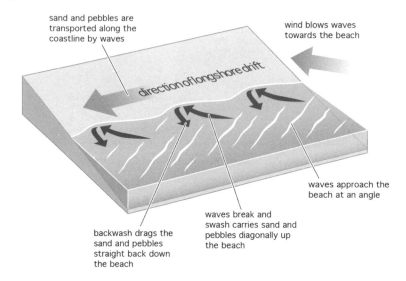

Material is deposited as the waves lose energy. This usually occurs in sheltered bays or where the coastline changes direction. In the same way as rivers, the largest material is deposited first.

The rate of coastal erosion depends on the type of wave and the geology of the area. If there are bands of alternating hard and soft rock the soft rock is eroded more quickly, leading to a pattern of **headlands** and **bays**.

Revision tip

The best way to remember the different types of waves is to remember that constructive waves construct or build up the coast whilst destructive waves destroy the coast. They also have opposite characteristics so if you learn the wave height, nature of the swash, backwash, etc. for one, you can easily work it out the other.

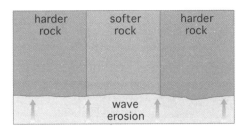

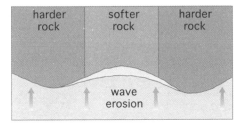

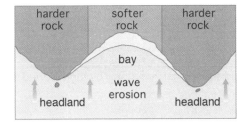

At a headland, caves, arches, stacks and stumps may form. Wave erosion causes cracks or faults to be widened into caves and then arches. Eventually the roof of the arch collapses to form a stack. This is further eroded to create a stump.

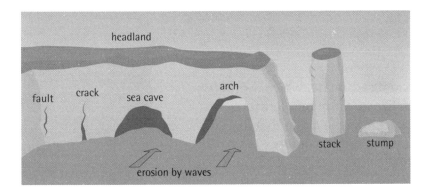

Cliffs are steep outcrops of rock along a coastline. Wave erosion creates a wave-cut notch at the base of the cliff. Over time the notch becomes larger and undercuts the cliff above. Eventually the cliff collapses. A wave-cut platform is left as the cliff retreats inland.

Activity

Create a PowerPoint presentation or a poster to show the different landforms that can be found at a coast. You could use annotated photos, maps and diagrams to explain how the landforms are formed and to give examples of where they may be located.

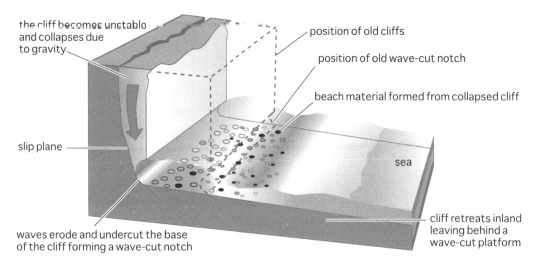

the cliff becomes unstable and collapses due to gravity

position of old cliffs

position of old wave-cut notch

beach material formed from collapsed cliff

slip plane

sea

waves erode and undercut the base of the cliff forming a wave-cut notch

cliff retreats inland leaving behind a wave-cut platform

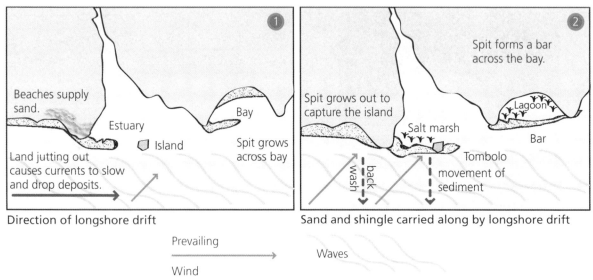

1

Beaches supply sand.

Estuary

Bay

Island

Spit grows across bay

Land jutting out causes currents to slow and drop deposits.

Direction of longshore drift

2

Spit forms a bar across the bay.

Spit grows out to capture the island

Lagoon

Salt marsh

Bar

back wash

Tombolo
movement of sediment

Sand and shingle carried along by longshore drift

Prevailing
Wind

Waves

Spits, bars and tombolos are all examples of landforms formed by coastal deposition. Where the coastline changes direction, sand and shingle are deposited. Sand dunes are also found at the coast where windblown deposits are held in place by marram grass.

How are coral reefs and mangrove swamps formed?

Coral reefs are made of limestone or calcium carbonate, which is produced by **polyps** (tiny marine organisms). Polyps live in warm temperatures (26–27°C) and in shallow, clear moving water. It is the algae that live alongside the polyps which give a coral reef its bright colours – coral itself is white. Coral is sensitive to environmental change and so conditions need to remain the same throughout the year. Most coral reefs occur between 30°N and 30°S.

The different types of reef are:

- Fringing reefs – found near shorelines. They grow in areas that are protected by a larger barrier reef.
- Barrier reefs – found further from the shoreline. They are often separated from the land by a **lagoon**.
- Atoll reefs – look similar to barrier reefs but are found around submerged volcanic, oceanic islands.

Mangrove swamps occur in the same regions as coral reefs as they grow in similar environments. They grow in sheltered areas and on gently sloping mud flats and can survive in both salt water and fresh water. Mangroves have stilt roots which provide support and absorb oxygen. This is important because mud flats have densely packed particles that cannot hold enough oxygen for trees to survive.

What opportunities and hazards do coasts present?

Opportunities presented by coasts include leisure and tourism. For example, people may enjoy walking on beaches or sunbathing or they may enjoy diving in coral reefs. There are also opportunities for fishing (particularly shrimps) in mangroves.

However, climate change is likely to present a significant threat. Coastal flooding due to storm surges, inland flooding, erosion and cliff collapse may threaten people's homes and livelihoods, particularly in areas susceptible to tropical storms. **Coral bleaching** may occur in reefs if environmental conditions change. Other threats include **overfishing**, shipping and pollution.

How can coastal erosion be managed?

Many coastlines are managed by National Parks or councils. Most aim to put in place strategies that are sustainable. However, often these can be expensive or have unintended consequences on other parts of the coastline. **Sea walls** or **groynes** can be built to reduce cliff erosion. **Soft engineering** strategies include **beach nourishment** and **dune stabilisation**. However, in tropical areas where coral reefs and mangroves are located, the greatest threat is climate change. Climate change is a global issue and is therefore difficult to manage at a local level.

What case studies do I need?

You need case studies on:

- The opportunities presented by an area or areas of coastline, the associated hazards and their management, e.g. the Great Barrier Reef, Australia.

Quick test

1. Sketch a diagram to show the development of headlands and bays.
2. Describe how a stump is formed.
3. What is the difference between a barrier reef and an atoll?
4. Define what is meant by 'coral bleaching'.
5. Give an example of a hard engineering management strategy.

Revision tip

Make sure that you learn the specific opportunities and hazards for the case study that you have chosen. Coastlines have very different characteristics depending upon where they are located. General descriptions do not have the depth and detail expected in a quality answer.

Activity

Make a list of opportunities that are found in your located case study. Now explain what would happen to each opportunity if a major hazard, such as climate change or coastal flooding occurred. What would be the 'knock-on' effects?

Revision tip

It is important to remember that whilst management may appear to be a sustainable solution in one place, it is likely to have unintended consequences elsewhere. For example, a sea wall built to protect one part of the coastline will shift the wave energy further down the coast. This area will experience more severe erosion as a result.

Activity

For the case study you have studied, describe how the coastline has been managed. Do you think the strategies are sustainable? Would you have done anything differently if you were in charge? Why?

2.4 Weather

What are the characteristics of a Stevenson Screen?

The **weather** is the state of the atmosphere at a given time. A weather station can be used to observe, measure and record the weather. **Meteorologists** often use digital weather stations as well as aerial and satellite images. The basic equipment required for a weather station is kept inside or close to a **Stevenson Screen**. These are located in an open area and at least 1.5 metres above the ground to allow the free flow of air. They are also painted white to reflect sunlight.

Shields instruments from rain, wind, sun

Maximum-minimum thermometer

Barometer

Rain gauge

Cabinet may house: dry-bulb thermometer, barometer, hygrometer, maximum-minimum thermometer

Hygrometer

Located in open area, away from buildings and tall vegetation

Painted white to reflect sunlight

Anemometer

Cabinet at least 1.5 metres above the ground, and slatted, to allow free flow of air

Door faces away from direct sunlight

Wind vane

> You must be able to:
> - Describe how weather data is collected
> - Make calculations using information from weather instruments
> - Use and interpret graphs and other diagrams showing weather and climate data.

Revision tip

Many schools have weather stations which are either digital or housed in a Stevenson Screen. If you can, try to have a look at one (but make sure you seek permission first). Have a look at the instruments – can you work out what each one measures?

Activity

There are six different pieces of equipment found in a typical weather station such as the one shown in the picture. Ask a friend to cover up one of the instruments in the picture without telling you which one they have chosen. Can you work out which one has been covered up?

How can temperature, sunshine and cloud be measured?

Temperature is measured using a **maximum-minimum thermometer**. This records the highest and lowest temperatures in a 24-hour period. It is kept in the Stevenson Screen so that the air temperature (rather than the temperature in direct sunlight) is recorded.

Sunshine is measured using a **sunshine recorder**. This is a glass ball mounted on a frame. As the sun shines, it burns the paper in the frame. The length of the burnt line shows how long the sun has been shining.

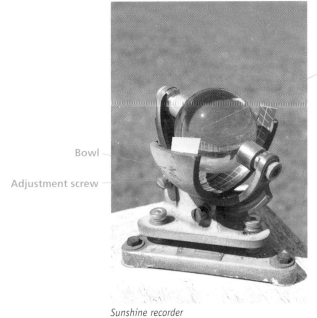

Glass sphere

Bowl

Adjustment screw

Sunshine recorder

Revision tip

Always check that the data you are presented with and work on seems sensible.

Activity

Find photos of each of the following cloud types to create a 'cloudspotters guide':

- Altocumulus
- Altostratus
- Cirrocumulus
- Cirrostratus
- Cirrus
- Cumulus
- Nimbostratus
- Stratocumulus
- Stratus

Meteorologists record both **cloud** type and cloud cover. There are three main types of cloud: cirrus, stratus and cumulus. Cloud cover is estimated and is worked out by how many eighths (oktas) of the sky are obscured by cloud. This is shown by shading segments of a circle.

 0 oktas (no cloud)   1 okta   2 oktas   3 oktas   4 oktas

 5 oktas   6 oktas   7 oktas   8 oktas (no blue sky)   Sky obscured

Activity

Have a look at this data set. Does it seem sensible? What might have happened?

Day	Time	Maximum Temperature (°C)	Minimum Temperature (°C)	Current temperature (°C)
1	9am	14	8	6
2	9am	10	−2	0
3	9am	11	−4	3

How can rain and relative humidity be measured?

Rain is measured using a rain gauge. A rain gauge captures the rain in a container, which can then be measured. It is important that this is located away from trees and buildings so that water dropping from them does not affect the data.

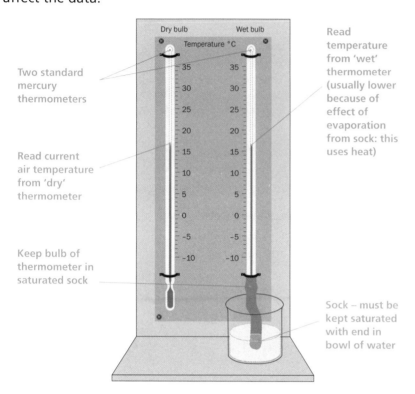

Two standard mercury thermometers

Read current air temperature from 'dry' thermometer

Keep bulb of thermometer in saturated sock

Read temperature from 'wet' thermometer (usually lower because of effect of evaporation from sock: this uses heat)

Sock – must be kept saturated with end in bowl of water

> ### Revision tip

In order to work out the relative humidity from the wet bulb and the dry bulb thermometer readings you will need to look at a table of figures. You can find an example here: http://mrsdlovesscience.com/realtivehumidity/realativehumidity.html. Alternatively you can plug your readings into the calculator here: http://www.ringbell.co.uk/info/humid.htm.

A **hygrometer** is used to calculate the **relative humidity** of the air. It contains two thermometers; a wet bulb and a dry bulb. The greater the difference between the two readings on the thermometers, the drier the air and the lower the relative humidity. The closer the humidity is to 100 per cent, the more likely it is to rain.

> ## Activity

Look at the hygrometer readings below and use a table or calculator to complete the relative humidity column. Then answer the questions that follow.

Reading	Wet bulb thermometer	Dry bulb thermometer	Relative humidity
1	20	24	
2	–3	–2	
3	5	16	
4	10	10	
5	23	30	

1. Which reading shows a relative humidity of 100%?
2. Which reading has the lowest relative humidity?

How can air pressure and wind be measured?

Air pressure is measured using a **barometer**. Air pressure is the force put on the Earth's surface by the weight of the air above it. Temperature is the most important factor determining air pressure. As air warms it rises, which causes low pressure. Rising air cools, becomes more dense, then sinks, causing high pressure.

Differences in air pressure cause air to move from areas of high to low pressure. This air movement is wind. Meteorologists record wind direction using a **wind vane** and wind speed using an **anemometer**. Wind direction can change frequently and so wind direction is recorded hourly. Anemometers convert wind speed to kilometres per hour (km/hr). However, wind strength can also be estimated using the **Beaufort Scale**.

What case studies do I need?

You need case studies on:

None for this unit. The focus is much more on being able to make calculations using information from weather instruments and on interpreting graphs and diagrams.

Revision tip

There is quite a lot of complicated vocabulary associated with this unit. Make sure you know all of the different words and also how to spell them.

Activity

Draw a diagram to show how air pressure works. Make sure you use colours to show the temperature of the air. Use the text here to help you.

Quick test

1. What is a Stevenson Screen?
2. Explain how sunshine is recorded.
3. What does a hygrometer do?
4. Which is more accurate – an anemometer or the Beaufort Scale?
5. What is measured in oktas?

How does the equatorial climate affect vegetation in the tropical rainforest?

The equatorial climate occurs along the Equator and 5° north and south of it. It has high temperatures (27°C) all year, with no clear seasons. Rain is frequent and heavy (2000mm per year). There is also high relative humidity, light winds and a very small annual temperature range (2 or 3°C).

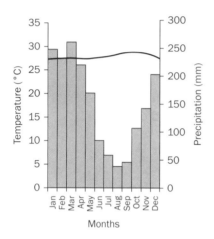

The climate is like this because of the latitude, altitude, air pressure and **diurnal temperature range**.

- **Latitude** – at the Equator the sun is high in the sky all year. **Insolation** is high and so temperatures are also high.
- **Altitude** – air temperature decreases with altitude. Most of the equatorial region is lowland and so temperatures are warmer.
- **Air pressure** – High temperatures mean that the air warms, becomes lighter and rises. This leads to low pressure, which brings heavy, **convectional rainfall**.

Tropical rainforest vegetation grows in equatorial regions. Climate conditions are ideal for plant growth and so there is high **biodiversity**. The high temperatures and rainfall weather the bedrock to create deep soils. However, the high rainfall results in the **leaching** of nutrients. Without a supply of **leaf litter**, soils become infertile.

You must be able to:
- Describe and explain the characteristics of equatorial and hot desert climates
- Describe and explain the characteristics of tropical rainforest and hot desert ecosystems
- Describe the causes and effects of deforestation of tropical rainforest.

Revision tip

It is useful to be able to sketch a climate graph of the equatorial climate. It does not need to have all of the figures on it but you can add detail by including annotations of the main climate characteristics.

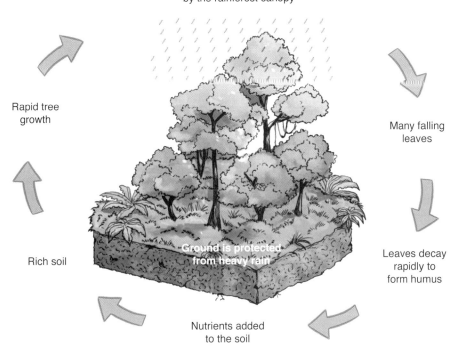

Heavy rain each day is intercepted
by the rainforest canopy

Rapid tree
growth

Many falling
leaves

Rich soil

Ground is protected
from heavy rain

Leaves decay
rapidly to
form humus

Nutrients added
to the soil

Activity

Look at the soil profile diagram below.

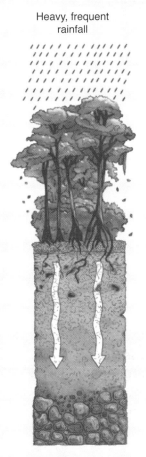

Heavy, frequent
rainfall

Add suitable annotations to the diagram to show the characteristics of the soil, vegetation and nutrient cycling in the tropical rainforest. You could also show the effects on the soil if deforestation occurs.

What are the causes and effects of deforestation of tropical rainforest?

Deforestation is occurring in the world's tropical rainforest at a dramatic rate. Logging has occurred as the global demand for hardwood, plywood and paper pulp has increased. Palm oil production has also increased and many local farmers start fires to clear forest. This creates farmland to feed the growing population.

If deforestation occurs, then there is no tree canopy to intercept rainfall. This then increases the rate of leaching and reduces the fertility of the soil further. The **habitats** of millions of insects and animals are also destroyed, meaning they are unable to survive. On a global scale, deforestation drives climate change. Trees absorb and store carbon. Deforestation releases this in the form of carbon dioxide, an important **greenhouse gas**. Deforestation has many negative effects on people. Desertification may occur and infertile soils make it difficult to grow crops effectively. Sediment may also be washed into rivers, polluting the water supply.

How does the hot desert climate affect vegetation in the hot desert ecosystem?

Hot deserts are found between 5° and 20° north and south of the Equator. They are characterised by a lack of rainfall – less than 250mm per year – and high temperatures – often above 30°C. Although temperatures are high all year round, there are distinct seasons. One that is cooler and wetter and one that is hotter and more **arid**.

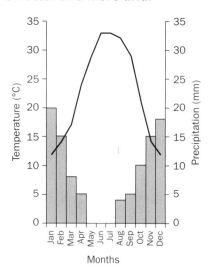

Hot deserts are areas of high atmospheric pressure. Due to the lack of cloud cover there is little rainfall and temperatures are intense during the day and fall rapidly at night. Therefore, diurnal temperature ranges are much greater than the equatorial region. Deserts can also be located on the leeward (sheltered) side of a mountain range. Here, the moisture in the air has already been lost as the air rises, cools and rains over the mountain.

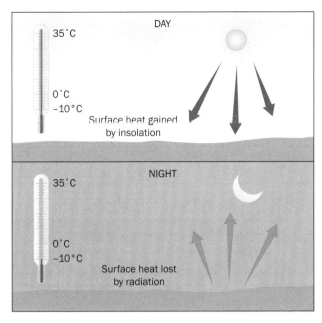

The biodiversity of hot deserts is very low. Whilst the high temperatures, long hours of daylight and unbroken sunshine are all good for plant growth, rainfall is light, unreliable and unpredictable. Soils are baked hard, which makes infiltration difficult. This, combined with high evapotranspiration rates, means there is little water available for plant growth. Desert soils are thin with little organic material and strong winds result in soil erosion.

Plants have adapted in a variety of ways:

- Dormancy – drought-resistant seeds lie dormant until a period of rainfall. These plants can complete their lifecycle within a few weeks.
- Water retention – some plants store water in stems, trunks or leaves.
- Tolerance of saline conditions – desert soils are salty because evaporation draws salt upwards towards the surface. Salt is toxic to plants but some have developed salt tolerance.
- Changes to roots systems and leaves – deep roots can reach groundwater, wide spreading roots catch sparse rainfall and small waxy leaves or spines reduce water loss.

What case studies do I need?

You need case studies on:

- An area of tropical rainforest, e.g. Borneo.
- An area of hot desert, e.g. the Namib Desert.

Quick test

1. Describe the climate in tropical rainforests.
2. Define what is meant by leaching.
3. Describe the climate in the desert.
4. Why is biodiversity in deserts low?
5. Suggest a way in which plants have adapted to desert conditions.

Revision tip

Remember that deserts are not always hot. There is minimal cloud cover (this explains the lack of rainfall), which means the sun shines brightly during the day and temperatures are high. However, it gets bitterly cold at night, sometimes below freezing, as the heat escapes quickly with little cloud cover. This means that the diurnal temperature range – the difference between the highest temperatures during the day and the coldest temperatures during the night – is huge!

Activity

Find a photograph of a plant or animal found in desert environments. Annotate the photograph to show how the plant or animal has adapted to the desert conditions.

Revision tip

Use the acronym DUST to remember some of the things that are mined in deserts such as the Namib. DUST stands for Diamonds, Uranium, Salt and then either Tin or Tungsten.

Activity

What might some of the impacts of human activity in desert environments be? Carry out some research for your case study and display your findings in a poster or presentation.

1　This photograph shows a coral reef which has undergone bleaching. Define what is meant by 'coral bleaching'.

....

....

.... [1]

2　Look at the data in the table below.

Day	Maximum temperature	Minimum temperature	Temperature range
1	22°C	4°C	
2	17°C	4°C	
3	10°C	–1°C	
4	13°C	2°C	

Calculate the range of temperature recorded on each of the four days.

[2]

3 Plate boundaries are where two parts of the Earth's crust meet. Describe how conservative plate boundaries result in earthquake activity.

....

....

....

....

....

....

....

....

....

....

....

.... **[3]**

4 Longshore drift is the main process of transportation along a coast. Draw a diagram which illustrates the process of longshore drift.

[3]

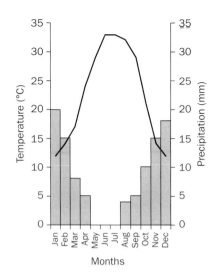

5 This climate graph shows the temperature and rainfall experienced in a desert. Describe the climate of a desert.

....

....

....

....

....

....

....

....

.... **[4]**

6 For a volcanic eruption you have studied, describe the social and economic impacts of the event.

....

....

....

....

....

....

....

....

....

.... **[5]**

7   For a river you have studied, explain how rivers can be managed.

....

....

....

....

....

....

....

....

....

....

....

....

....   [7]

3.1 Development

How is development measured and why do inequalities exist?

Development is the improvement in people's quality of life. It can be linked to the wealth of a country. Gross Domestic Product (GDP) per capita is the amount of money earned per person on average. It is an indicator of development and can be used to give an idea of how rich countries are. It does not take into account other aspects such as education, healthcare and access to clean water, which are also useful **indicators of development** and are essential for a good quality of life.

The **Human Development Index (HDI)** describes social and economic well-being. It uses adult literacy, life expectancy and GDP per capita to give a score between 0 and 1. Countries with a score close to 0 have low human development whilst countries with a score close to 1 have high human development.

Because lots of different indicators can be used to measure development and the development of countries changes over time, categorisation is difficult. In the 1980s countries were divided into **More Economically Developed Countries (MEDCs)**, which were generally located in the northern hemisphere and **Less Economically Developed Countries (LEDC)**, which were generally located in the southern hemisphere. Since then other categories have been identified that include **Least Developed Countries (LDCs)**, **Newly Industrialising Countries (NICs)** and **Brazil, Russia, India and China (BRICs)**.

There are a number of factors that affect the level of development of a country. These factors will be more or less important in different areas and countries and this has led to inequality within and between countries.

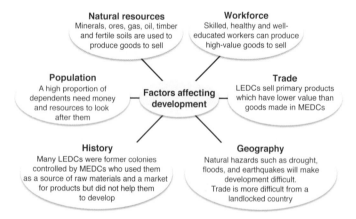

Development indicators are average figures for the whole population. They do not show the differences between rich and poor within a country.

In 2015 the UN agreed 17 **Sustainable Development Goals (SDGs)** aimed to reduce inequality between countries.

You must be able to:

- Use a variety of indicators to assess the level of development of a country
- Identify and explain inequalities between and within countries
- Classify production into different sectors and give illustrations of each
- Describe and explain how the proportions employed in each sector vary according to the level of development
- Describe and explain the process of globalisation and consider its impacts.

Revision tip

There are a lot of acronyms to learn in this section. It is worth learning them and also making sure that you have at least one example of a country that fits with each acronym. There are also some common development indicators that you should know. For example, infant mortality, literacy and life expectancy.

Activity

Carry out some research about the Human Development Index. Which countries are at the top? Which countries are near the bottom? Why do you think this is?

How is production classified?

Economic activity describes the ways in which people earn a living. All economic activities can be divided into four sectors: primary, secondary, tertiary and quaternary industries.

- Primary – industries that take something out of the earth. They are also known as extractive industries. Some primary industries produce goods that are sold exactly as they are. Others sell **raw materials** to make different products. Examples include fishing, farming, mining, quarrying and forestry.
- Secondary – industries that take raw materials and make new products out of them. They are also known as processing or manufacturing industries. Finished goods can be sold as **exports**. Examples include diesel, computers and clothes.
- Tertiary – industries that provide a service. Examples include hospitals, schools, financial services and tourism.
- Quaternary – industries that provide knowledge and information. Examples include research and development for high-tech companies such as mobile phone manufacturers.

Revision tip

It is important to learn the different sectors of economic activity and their characteristics. However, it is also important to have at least one example to illustrate each sector.

Activity

Think about what you want to do as a career (or, if you don't know, pick a career that you know something about). How would this career be classified? Does it link with or rely on other jobs in other sectors?

How do the proportions employed in each sector vary with the level of development?

As a country develops, the proportion employed in each sector changes. LEDCs have a high proportion of people employed in primary industries. As they develop, so a higher proportion of people are employed in secondary and tertiary industries.

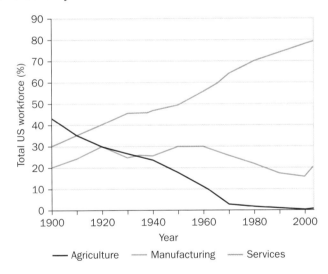

In some LEDCs, as jobs in the primary sector are lost there is not enough employment in the other sectors for those who need it. This leads to high unemployment and high underemployment. This has led to the growth of the **informal sector** which includes jobs that are not taxed or monitored. Examples include the selling of crafts and the provision of services such as shoe shining or car repair.

In contrast, some LEDCs have economies that are growing very quickly. Known as Newly Industrialising Countries (NICs), they have developed their secondary industries. The original NICs, the tiger economies of South Korea, Singapore, Taiwan and Hong Kong, developed in the 1970s. They are now considered MEDCs and their place has been taken by the BRIC economies.

Look at the table below. Sketch pie charts for each country to show the proportions of each type of industry. Can you use this information to work out which development category each country fits into?

Country	Primary	Secondary	Tertiary
Argentina	10.5%	29.1%	60.4%
Ivory Coast	17.4%	20.3%	62.2%
The Gambia	75%	19%	6%
Japan	2.9%	26.2%	70.9%
Russia	4.4%	35.8%	59.7%

The proportion of secondary and tertiary industries is higher in wealthy countries as these activities add more value than primary industries. As a general rule, the higher the percentage of primary industry, the poorer the country.

What is globalisation and what are its impacts?

Globalisation describes the increase in connections between places. These connections can be through migration, goods, services and money. Globalisation has occurred as a result of improvements in transport, the development of technology such as the internet and the removal of trade barriers. **Transnational corporations (TNCs)** have also grown. They have their headquarters in MEDCs but have moved production into LEDCs to reduce their costs.

The positive impacts of globalisation include:

- TNCs invest in countries by developing the local infrastructure and creating jobs.
- TNCs usually pay higher wages than local businesses. Employees therefore have a higher quality of life.
- TNCs bring wealth to the local economy, which has a positive **multiplier effect**.
- Local people learn new skills and use new technologies.
- Most manufactured products are exported, which benefits the economy.
- TNCs pay tax to the host country. This can be invested in health, education and infrastructure.

The negative impacts of globalisation include:

- TNCs have a reputation for making employees work long hours in poor conditions.
- Profits are often sent back to MEDCs where the TNC has its headquarters.
- TNCs can transfer production to other low-cost locations with little warning.
- There are fewer laws about pollution in many LEDCs so TNCs may damage the environment without being prosecuted.
- LEDCs may become too dependent on TNCs.

There are lots of advantages and disadvantages listed here. Try to learn at least two of each and make sure they relate to your case study.

What case studies do I need?

You need case studies on:

- A transnational corporation and its global links, e.g. Nike.

Quick test

1. What is the HDI?
2. Give an example of a tertiary industry.
3. What is meant by the informal sector?
4. Give an example of a TNC.
5. Suggest one advantage of TNCs.

Activity

Have a debate with a friend about the advantages and disadvantages of TNCs and globalisation. Do you think that overall they are a good thing or a bad thing?

3.2 Food production

What are the main features of agricultural systems?

Farming is a type of primary industry. There are many different types of farming and they can be classified in three main ways:

- By what we get out of them – **commercial** or **subsistence**
- By what we put into them – **intensive** or **extensive**
- By what is grown – **arable**, **pastoral** or **mixed**

In addition, **nomads** are farmers who move from one place to another whereas **sedentary** farming occurs in one place.

Farming can be described using a systems diagram with **inputs**, **processes** and **outputs**.

- Inputs – the things that go into the system, e.g. climate, soils, machinery, animals, money.
- Processes – the things that happen to the inputs to turn them into outputs, e.g. ploughing, planting, feeding, calving.
- Outputs – the things that come out of the system, e.g. wheat, vegetables, pork, chicken, milk, wool, money (profit).

Inputs can be further classified into natural (physical) and human inputs. Crops need fertile soils and a mild climate in which to grow. In some areas of low rainfall, **irrigation** can be used although it can be expensive. Steep slopes are difficult to farm as their soils are thinner and it is harder to use machinery. Farms also need enough **labour** to work on the farm. In some places labour may be seasonal. Farms that have a lot of money and technology and are efficient and make big profits are known as **agribusinesses**.

> **You must be able to:**
> - Describe and explain the main features of an agricultural system: inputs, processes and outputs
> - Recognise the causes and effects of food shortages
> - Describe possible solutions to these problems.

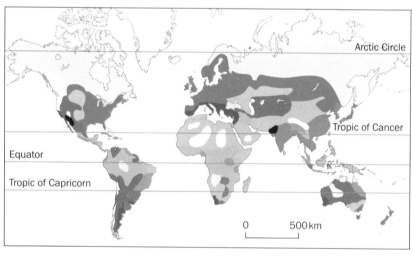

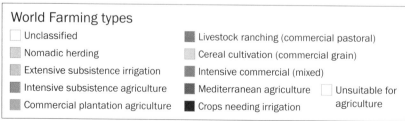

World Farming types

- Unclassified
- Nomadic herding
- Extensive subsistence irrigation
- Intensive subsistence agriculture
- Commercial plantation agriculture
- Livestock ranching (commercial pastoral)
- Cereal cultivation (commercial grain)
- Intensive commercial (mixed)
- Mediterranean agriculture
- Crops needing irrigation
- Unsuitable for agriculture

In tropical areas, much of the commercial agriculture is carried out on **plantations**. They often grow a single type of crop, such as palm oil, coffee or tobacco, for export; this is known as monoculture. Subsistence farming occurs in other parts of tropical areas. In the Amazon and Congo rainforests tribes engage in **shifting cultivation**. Farmers cultivate small plots of land near their village. As the soil fertility decreases, they move to farm a neighbouring area.

What are the causes of food shortages?

Food provides people with the **calories** that are needed to grow and develop healthily. The average number of calories needed per day is around 2500. It is estimated that there is enough food for everyone in the world to consume at least this amount but the food is not distributed evenly. Also, as population increases so food is in even shorter supply in some areas.

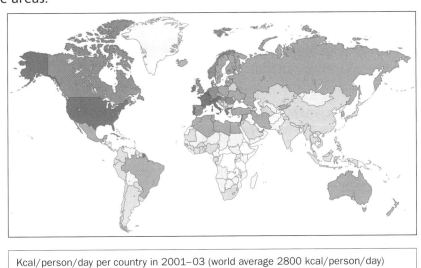

Kcal/person/day per country in 2001–03 (world average 2800 kcal/person/day)

☐ 1600–2000　　　☐ 2000–2400　　　☐ 2400–2800
▨ 2800–3200　　　▨ 3200–3600　　　■ More than 3600

Food prices rise as demand from wealthier countries, including India and China, increases. Poorer countries that do not grow enough of their own food can no longer afford to import enough for their populations. Other causes of food shortages include:

- Extreme weather events, e.g. floods and drought, that cause crops to fail.
- Low capital investment.
- Poor distribution and transport difficulties.
- Natural disasters, e.g. earthquakes, affect food distribution by damaging transport links.
- Outbreaks of disease mean people cannot work and grow food.
- Pests, e.g. locusts, destroy crops.
- Conflict, e.g. in Darfur and in Sierra Leone, disrupts food supplies.
- As populations have increased, unsuitable land is farmed, leading to soil erosion and soil exhaustion.

As the world's climate changes, so the United Nations has estimated that three-quarters of the world's poorest people will be affected by crop failures that are a result of global warming.

Revision tip

There are lots of terms to learn here. Play key term bingo with some friends. Each of you should select five of the terms and write them down. Someone should be the caller and read out definitions. The first person to tick off all of their key words is the winner.

Activity

Draw a systems diagram for a type of farming.

Revision tip

Population growth is a cause of food shortages but it is a much more complex problem than this. Make sure you understand and can explain this complexity.

Activity

Explain why global warming is likely to increase food shortages in the future.

What are the effects of food shortages?

A lack of food, vitamins or minerals can cause **malnutrition**. Some 13% of the world's population do not have enough food. Most of these people live in LEDCs and many are children. There are a number of diseases which can result from food shortages.

- Marasmus – sufferers become thin and stop growing.
- Kwashiorkor – the stomach swells, skin peels and hair turns orange.
- Anaemia – lack of iron.
- Vitamin A deficiency – causes blindness and death.
- Iodine deficiency – affects the brain and can be fatal.

What are possible solutions to the effects of food shortages?

When poor countries experience a food shortage often richer countries provide food aid. This does little to solve long-term problems as the country may become dependent on hand-outs. Increasingly, agricultural improvements have been suggested so poorer countries can maintain their food security.

- The Green Revolution – high yield varieties of rice, maize and wheat developed. Food production has increased as crops are more resistant and grow faster. More jobs are available and some farmers have become richer and employ more local people. However, the less well-off cannot compete. They may have borrowed money or sold their land and moved to the city. Chemicals have polluted water supplies and irrigation has increased the demand on drinking water stores.
- Genetically Modified (GM) crops – the altering of crops to withstand pests and diseases. They have potential to increase the amount of food that can be grown. However, some countries have banned GM crops until they are proved safe for people and the environment.
- Intermediate and appropriate technology – simple solutions that use existing local skills and are more affordable for poorer countries. For example, saving land from erosion by building terraces of gently sloping land.

Terraced fields in the Annapurna region of Nepal

What case studies do I need?

You need case studies on:

- A farm or agricultural system, e.g. plantation agriculture in Sarawak, Malaysia, or rice farming in Bangladesh.
- A country or region suffering from food shortages, e.g. food shortages in Bolivia.

Quick test

1. What is intensive farming?
2. Name an output of farming.
3. What is kwashiorkor?
4. What are GM crops?
5. Why is giving aid not a good solution to food shortages?

Revision tip

Make sure that you understand the disadvantages as well as the advantages of agricultural improvements such as the Green Revolution.

Activity

Carry out some research into the Green Revolution, GM crops and intermediate and appropriate technology. Which agricultural advance would you recommend to a poor community in order to improve their food security?

3.3 Industry

What are the main features of industrial systems?

Different types of industry include **manufacturing**, **processing**, and **assembly** and **high-tech industries**. These are all examples of **secondary sector** employment. As a country develops, its secondary industry becomes more important than its primary industry. It is therefore an important aspect of a country's development. In the same way as agriculture, industry can be viewed as a system. There are inputs, processes and outputs. There are also **feedbacks** where outputs, such as waste, are recycled or can be used as a source of energy to power factories.

> **You must be able to:**
> - Demonstrate an understanding of an industrial system: inputs, processes and outputs
> - Describe and explain the factors influencing the distribution and location of factories and industrial zones.

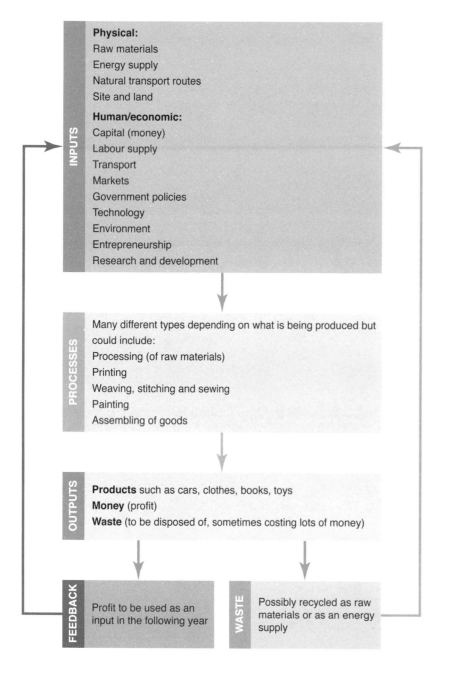

INPUTS

Physical:
Raw materials
Energy supply
Natural transport routes
Site and land

Human/economic:
Capital (money)
Labour supply
Transport
Markets
Government policies
Technology
Environment
Entrepreneurship
Research and development

PROCESSES

Many different types depending on what is being produced but could include:
Processing (of raw materials)
Printing
Weaving, stitching and sewing
Painting
Assembling of goods

OUTPUTS

Products such as cars, clothes, books, toys
Money (profit)
Waste (to be disposed of, sometimes costing lots of money)

FEEDBACK

Profit to be used as an input in the following year

WASTE

Possibly recycled as raw materials or as an energy supply

> **Revision tip**
>
> Processes are verbs, they are 'doing words' because a process involves doing something to the inputs. Generally, therefore, most processes end with 'ing'. Spot the processes in the list below.
>
> capital stitching assembling labour mobile phone waste recycling energy processing

> **Activity**
>
> Choose something that belongs to you that has been manufactured. Write a list of all of the inputs, processes and outputs that you think have gone into making the object you chose. Does the list surprise you?

What factors influence the distribution and location of factories and industrial zones?

There are a number of factors which influence where an industry is located.

- **Raw materials** – these can be bulky and cost a lot to transport. Some industries may need to be close to their raw materials because of this.
- Energy supply – in some countries without a national electricity supply it is important for factories to be located close to a power station.
- Natural transport routes – mountains make transport difficult whilst valleys make transport easy.
- **Site** and land – many factories need large areas of cheap, flat land to build on.

An industrial area on the outskirts of a town

- Capital (money) – big factories cost a lot of money and so rich investors may be needed.
- Labour supply – factories need skilled and / or unskilled workers.
- Transport – ports, railways and good roads make transport of raw materials and finished goods faster and easier.
- Markets – large numbers of people living close by to purchase products will help businesses to thrive.
- Government policies – governments can encourage industry by offering **grants** and tax breaks.
- Technology – e-mail and other forms of communication mean that factories can now be built a long way from headquarters.
- Environment – a pleasant landscape and good local facilities may attract workers.
- **Entrepreneurship** – businesspeople may want to set up new companies near where they live or where they were born.
- Research and development (R & D) – industries that rely on advanced technology need to locate in countries where there are scientists conducting relevant research.

Revision tip

The factors that are important to an industrial location will depend hugely upon the industry itself. Make sure you know the factors that are important to your particular case study and learn them.

Activity

Find a map of your local area. Either mark where you think the best place to locate a factory would be and annotate the map to explain your suggestion; or if there is a factory located there already, explain why this is a suitable location. Use the factors listed here to help you write your annotations.

Raw materials and a power supply were traditionally the most important factors affecting industrial location. Transport was difficult and expensive so factories would locate close to bulky raw materials. Power was generated through water, wood or coal. At the beginning of the 20th century transport became easier and so raw materials were more available. Power was produced by power stations connected to a network. The most important factor became the cost of labour.

A good example of how industrial location has changed over time is steel. In the 19th century factories located close to heavy iron ore and limestone deposits in the UK. Today, the industry is located in China and India as labour is cheaper than that in the UK.

High-tech industries are those that manufacture technology such as computers, mobile phones, robotics and computer consoles. These products are small and light and can be made anywhere in the world. They are known as **footloose** industries. The R & D needed for high-tech industries means they need to attract **graduates**. High-tech industries are therefore located in attractive environments where people will want to work.

Clusters of high-tech industries are often located in **science parks**. They are located close to universities and have excellent transport and communication connections. Often, large amounts of capital are needed to buy land and specialist laboratories. For these reasons, science parks are often located in MEDCs and NICs. For example, the Hsinchu science park in Taiwan has over 400 high-tech industries. National Chiao Tung University and National Tsing Hua University are located nearby. The park has strictly monitored air quality and prides itself on being environmentally friendly.

Revision tip

You should understand that the importance of factors of industrial location change over time. How might the factors in your case study have changed? How might they continue to change in the future?

Activity

Explain why, today, many toys and clothes are made in Taiwan and China.

Revision tip

Remember that high-tech industries and their associated science parks and research and development are fairly recent innovations. Quaternary industry did not really exist before the middle of the last century and does not feature in the employment profile of many, particularly poorer, countries.

Activity

Play hangman with a friend using the key words and factors of industrial location from this unit.

The Hong Kong Science Park

What case studies do I need?

You need case studies on:

- An industrial zone or factory, e.g. the textile and clothing industry in Bangladesh.

Quick test

1. Give an example of a feedback in an industrial system.
2. How does entrepreneurship affect industrial location?
3. What were the two factors affecting industrial location in the 19th century?
4. What does footloose mean?
5. What is a science park?

3.4 Tourism

Why has the tourist industry grown?

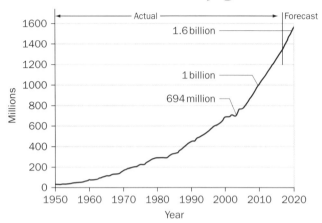

Tourism has grown dramatically over the last 50 years. Different parts of the world have seen growth take place at different rates. This growth has taken place due to:

- Rapid air travel has taken over from slow sea journeys.
- The development of efficient planes has made air travel cheaper.
- Transport facilities have improved so journey times have been cut.
- Wealth has increased in many countries meaning people have more **disposable income** to spend on luxuries.
- As people have become richer they have demanded more paid holiday time.
- The internet, advertising and holiday programmes have meant people have a wider knowledge and desire to visit other countries.
- People cope with the stress of modern life by going on holiday.

People are often attracted to holidays in places which have different human and physical landscapes to the places where they live. For example, tourists from the UK tend to visit places with beaches and warm weather. Types of holidays have also changed over time. Whilst **package holidays** used to be common, there are now a wide variety of holidays to choose from. For example, **adventure tourism**, **wilderness tours** and **ecotourism**.

The type of tourist has also changed. For many years most tourists came from Europe and the USA. As NICs and LEDCs have become richer so more people from these countries are able to afford to go on holiday. People are now travelling further afield. For example, whilst many Chinese tourists visit nearby countries such as Thailand and South Korea, some are now travelling to major world cities such as Paris, Rome and London.

What are the advantages of tourism?

Tourism can bring vast amounts of money into the economy. In 2009 the global tourist industry was worth $852 billion. It also brings local employment opportunities. The money that workers earn can be spent in their local community so there is a **multiplier effect** as knock-on benefits reach shopkeepers and other businesspeople. A positive international image may also lead to foreign investment from foreign countries, which is particularly important for LEDCs. In order for tourism to be a success

You must be able to:
- Describe and explain the growth of tourism in relation to the main attractions of the physical and human landscape
- Evaluate the benefits and disadvantages of tourism to receiving areas
- Demonstrate an understanding that careful management of tourism is required in order for it to be sustainable.

Revision tip

Remember that when describing graphs you need to firstly, describe the general trend, then give figures to support the general trend and finally, suggest if there are any anomalies. Go through these three stages to describe the graph shown here.

Activity

Extend the lines on the graph to predict what you think will happen to the number of tourists in the future. Why do you think this will happen?

Revision tip

Try to remember three or four advantages of tourism. It's helpful to use your own experience to remember some of the key points in this section. Think about where you go on holiday. What advantages does tourism bring to the place you go on holiday? What might this place be like if tourism wasn't a function?

Activity

Rank the advantages of tourism mentioned here. Which do you think is most important? Which do you think is least important? Why do you think this?

it needs good **infrastructure**. Governments therefore may spend money on infrastructure and services, such as swimming pools, that benefit local people as well as tourists.

Many tourists visit cultural attractions and so money is made available to keep these places in good condition. Also, some visitors want to enjoy beautiful scenery and so this means that the environment needs to be protected.

What are the disadvantages of tourism?

Whilst tourism does provide employment, much of this is **seasonal** and poorly paid. There are many opportunities during the summer but in many places everything shuts down in the winter and there is not much for people to do. There can also be conflict between the tourists and the locals. Vast numbers of tourists can cause congestion and traffic problems, whilst pollution and waste may increase as a result of tourist developments. Tourists may also not be sensitive to local customs and may dress inappropriately.

Whilst a large amount of money may be invested in areas tourists are attracted to, other neighbouring areas may miss out. Another problem is that tourist developments may be owned by foreign companies and so profits may be taken overseas. Land that could have been used for building houses or growing crops is used instead for tourism. Crime may increase and the culture of a country may become increasingly **westernised**. On a global scale, the number of people travelling by air is contributing to climate change.

Why is the management of tourism required in order for it to be sustainable?

There is a fine balance between the advantages and disadvantages of tourism and so **management** is needed in order for tourism to be **sustainable**. Various groups, including National Park authorities, governments, conservationists and tourist companies are responsible for developing management strategies. Depending on the nature of the tourism this may involve raising awareness of issues through posters or developing public transport so traffic congestion is less of a problem. Tourism is more likely to be sustainable if local people are involved in decision making and management.

What case studies do I need?

You need case studies on:

• An area where tourism is important, e.g. France or Kenya.

Quick test

1. What has happened to the number of tourists globally over the last 50 years? (See graph on page 76).
2. Name **one** type of holiday.
3. What is meant by 'disposable income'?
4. Why might seasonal employment be a problem?
5. Suggest a tourism management strategy and explain why it is sustainable.

Revision tip

When writing about the advantages and disadvantages of tourism remember to use tentative language. For example, don't write that 'all employment related to tourism is seasonal and poorly paid' – it isn't! Instead, write 'sometimes' or 'generally work related to tourism is seasonal and poorly paid'.

Activity

Draw a cartoon strip to show the disadvantages that tourism can bring to an area. Use speech bubbles and captions to explain what the issue is.

Revision tip

Make sure that your case study covers this aspect in detail and that you can describe the management strategies clearly.

Activity

For all of the disadvantages that you have drawn on your cartoon, suggest ways in which these could be managed. You might want to draw a second cartoon which shows these strategies. Are there any problems that you think will be impossible to manage effectively? Why?

3.5 Energy

What are the different types of energy?

Energy can be divided into two types: non-renewable and renewable. **Non-renewable energy** comes from sources that cannot be replaced. Examples include **fossil fuels** such as oil, gas and coal. **Renewable energy** can be used over and over again without running out. Examples include wind, water and solar power.

Non-renewable energy

- Oil – oil-fired power stations are usually built on coastlines. This is because oil pipelines can easily bring the liquid ashore or tankers can bring it from other countries. The estuaries need to be deep so the tankers can come in to port. In some countries, oil-fired power stations have to be built away from people because of pollution and safety risks.
- Gas – natural gas is often found under the sea in the same place as oil and is piped ashore in a similar way. Some countries have onshore gas. This is transported using networks of pipes to power stations.
- Coal – coal is a heavy, bulky raw material. It is expensive to transport. Power stations are usually found close to coal mines and near to industrial towns which use the energy the power stations provide.

Renewable energy

- Geothermal – countries with active volcanoes use energy from heated rocks and magma. In Iceland geothermal power stations produce a quarter of the country's electricity and 90% of heating and hot water.
- Wind – wind farms can be located on land or offshore and need a good supply of wind. On land, flat areas are needed to build wind farms on. Wind farms are expensive to build but are cheap and safe to operate. They are almost pollution-free.
- Water – the most widely used form of water power is hydro-electric power. Large, fast flowing rivers with a steep gradient are needed to generate this type of energy. France has tidal barrages (dams) across the mouths of some of its big rivers to generate tidal power.
- Solar – solar panels convert heat from the sun to electricity. Energy can be generated on a small scale by installing solar panels onto people's houses.
- Biogas – rotting fruit, vegetables and animal waste give off methane. This can be burnt as biogas. This has been successful on a small scale in LEDCs. Some European countries are using the technology on a larger scale by connecting biogas plants into natural gas pipelines.

You must be able to:
- Describe the importance of non-renewable fossil fuels, renewable energy supplies, nuclear power and fuelwood; globally and in different countries at different levels of development
- Evaluate the benefits and disadvantages of nuclear power and renewable energy sources.

Revision tip

Fossil fuels – oil, gas and coal – are made out of fossilised plants and organisms. Arguably, they are being replaced but at such a slow rate that they are classified as non-renewable. For example, it takes coal about 300 million years to form!

Activity

Research the **energy mix** for the country that you live in. Why do you think these sources of energy are used?

How does energy consumption change over time and between countries?

Over the last 200 years the amount of global energy **consumption** has increased dramatically. This is due to the rapid increase in global population and an increase in gadgets which use energy. Most of this energy comes from non-renewable fossil fuels. However, the use of renewable energy is increasing. Whilst new coal and oil fields are continuously being discovered, they are getting harder and more expensive to find. It is difficult to keep up with the demand from MEDCs and industrial countries such as China and India.

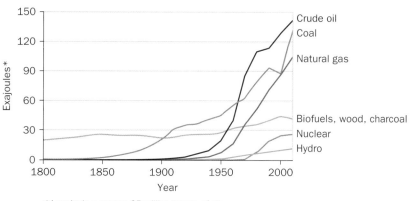

1 exajoule = approx. 25 million tonnes of oil

The amount of energy used around the world is not the same, nor is the energy mix. Levels of consumption are likely to depend on the population size and the level of development of a country. The energy mix will depend upon the sources of energy which are available and whether or not a country can afford to import energy from elsewhere. For example, in Saudi Arabia large amounts of energy are used, almost exclusively from oil. This is because Saudi Arabia has massive oil fields and therefore oil is very cheap.

Oil pumps working in an oil field

Revision tip

Make sure that you know the energy mix for your chosen case study and can quote some figures to illustrate this.

Activity

Look at the data in the table below, which shows energy sources as a percentage of the global total. Display this data in a graph or chart.

Energy source	Percentage
Oil	36
Natural gas	24
Coal	28
Nuclear	6
Hydro	6

What are the benefits and disadvantages of nuclear power?

Nuclear power uses uranium as its raw material. It is not renewable, but less is needed compared to fossil fuels. France, Lithuania, Slovakia and Belgium all use nuclear power to generate over 50% of their energy. Most nuclear power stations are found in MEDCs as the technology to develop nuclear power is expensive. Nuclear power stations need a nearby water supply for cooling the **nuclear reactor** and flat, cheap land. They also tend to be found away from towns and cities as there have been concerns over the safety of nuclear power. Despite this, it is a relatively clean, safe technology which is better for the environment than fossil fuels.

Advantages	Disadvantages
• The amount of raw material needed is very small compared with other fuels such as coal or oil. • It contributes relatively little to acid rain, global warming and climate change. • There is a lot of research being carried out to address fears over safety. • It receives support from many governments around the world. • Many measures are taken to make sure it is as safe as possible. • It allows countries to reduce their dependence on fossil fuels and to cut imports of these resources.	• Many people still have serious concerns over the safety of nuclear technology. • There is the potential for disastrous accidents to take place. Sometimes, as with the case of the Japanese tsunami, these accidents can be the result of natural disasters. • People living near to nuclear power plants may suffer from poor health. • Nuclear waste is difficult and expensive to dispose of and can be dangerous for many years. • Nuclear power plants are very expensive to build. • Nuclear power plants are very expensive to close down when they are no longer used.

Revision tip

In order to have a balanced argument you need to be able to give both advantages and disadvantages of nuclear power. Whilst your lists don't have to have exactly the same number of points, make sure they are not too lop-sided!

Activity

Carry out some research into events that have fuelled safety concerns over nuclear energy. For example, Chernobyl or the events following the 2011 Japanese earthquake (you might want to combine this with your case study for 2.1 Earthquakes and volcanoes).

What are the benefits and disadvantages of renewable energy?

Hydro-electric power (HEP) requires high volumes of water to drive the **turbines** that produce electricity. HEP stations are found at waterfalls or where water flows down a steep hillside. Major users of HEP include Iceland, Norway, Paraguay, China, Brazil and the Democratic Republic of Congo. For example, the Colorado River in the USA generates a huge amount of HEP. Its deep canyons and gorges make ideal sites for storing water and some have become tourist attractions. However, environmentalists are concerned about the lower rates of flow downstream, which have had negative effects on the wildlife.

Advantages	Disadvantages
• Power is cheap to produce and generally reliable. • HEP creates very little pollution or waste. • Dams reduce the risk of flooding. • Dams are often built in highland areas away from major population centres. • Reservoirs behind dams can be important sources of water. • Reservoirs are often used for leisure and tourism. • HEP stations can very quickly increase the amount of power they produce when required.	• Finding suitable sites can be difficult. • Dams and HEP stations are expensive to build. • Valuable areas of land may be flooded when dams are built. • Dam collapse can cause widespread damage. • There may be visual pollution and environmental damage to wildlife habitats. • Water quality and quantity may be reduced downstream, causing water shortages. • Large rivers can deposit a lot of very fine material which clogs up the turbines.

What case studies do I need?

You need case studies on:

- Energy supply in a country or area, e.g. coal supply in China.

Quick test

1. What is geothermal energy?
2. What is the most used energy source globally?
3. How has energy consumption changed over the last 200 years?
4. Suggest a disadvantage of nuclear power.
5. Suggest a disadvantage of hydro-electric power

 Revision tip

Whilst renewable energy may seem like a good alternative source of energy, it is not without its problems. Again, it is important to be aware of these problems so that you can produce a balanced argument.

 Activity

For another type of renewable energy, e.g. wind or solar power, write a list of its advantages and disadvantages.

3.6 Water

What are the different methods of water supply?

Clean water is available from a number of sources.

- Lakes and rivers – these are the main sources of accessible water. However, many lakes and rivers have been polluted by waste or are over-used.
- **Aquifers** – these are underground sources of water. Water is extracted via bore holes or wells. However, there needs to be a balance between the amount of water extracted and the amount that can be replaced naturally by rainwater.
- **Desalinisation** plants – these remove the salt from seawater. This can be a very expensive process and can only be afforded by wealthy countries.
- Large-scale rainwater harvesting – this is the collection and storing of rainwater in reservoirs which can be created by dams. Rainwater harvesting is seen as the best way of storing large quantities of water for the future.
- Small-scale rainwater harvesting – this is the collection and storage of rainwater by individual properties.

A rainwater collection tank

You must be able to:

- Describe methods of water supply and the proportions of water used for agriculture, domestic purposes and industrial purposes in countries at different levels of economic development
- Explain why there are water shortages in some areas and demonstrate that careful management is required to ensure future supplies.

Revision tip

As well as learning the different sources of water it is also useful to learn some of their characteristics. For example, that desalinisation is expensive or that large-scale rainwater harvesting is likely to be the best method of storing large quantities of water for the future.

Activity

Carry out some research to find out where your region or country gets its water from.

How do the proportions of water use vary according to purpose and economic development?

Water is used for a variety of purposes, not just for drinking, cleaning and cooking. **virtual water** is water that is used indirectly in the production of goods. For example, it takes 11 000 litres of water to make a pair of jeans and 15 600 litres of water to produce 1kg of beef.

The proportion of water used for agriculture, industry and domestically varies between countries and with the level of economic development. Generally, LEDCs use a higher proportion of water for agriculture than MEDCs. Also, MEDCs tend to use a higher proportion of water for industry and electricity generation than LEDCs. The amount used domestically varies dramatically and depends on a number of factors including the size of the population and the relative importance of other purposes.

Revision tip

The proportion of water used for different purposes is quite complicated. For example, crops such as rice, which are grown in flooded padi fields, will use more water than crops such as wheat or barley. Being able to explain this will make your answers more sophisticated.

Country	GDP per capita (US$)	National water use (av. litres per person per day)	National % water for agriculture	National % water for domestic activities	National % water for industrial activities and electricity generation
USA	46 000	575	41	13	46
UK	35 000	149	3	3	75
Russia	17 500	125	18	19	63
Mexico	14 300	366	78	17	5
Canada	38 000	483	12	68	20
France	34 000	287	10	16	74
China	7200	86	67	7	26
Egypt	6100	252	78	8	14
India	3300	135	86	8	6
Ghana	2100	36	48	37	15

Activity

Find out the proportions of water use in your country. Can you explain why they are like they are? You might want to compare the figures to the proportions of people employed in each sector.

Why are there water shortages in some areas?

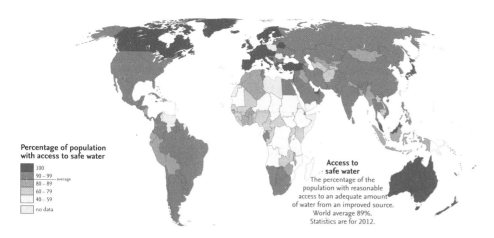

Percentage of population with access to safe water

- 100
- 90 – 99 — average
- 80 – 89
- 60 – 79
- 40 – 59
- no data

Access to safe water
The percentage of the population with reasonable access to an adequate amount of water from an improved source.
World average 89%.
Statistics are for 2012.

Many people have unlimited access to clean drinking water. However, there are over 1 billion people who struggle to get enough water to meet their **basic human needs**. This number is likely to increase to 4 billion people by 2050. There is not a lack of water but the main problem is accessibility. This is because rain does not always fall where it is most needed.

Revision tip

It is important to remember that there is enough water on the planet; the problem is that it is not distributed equally.

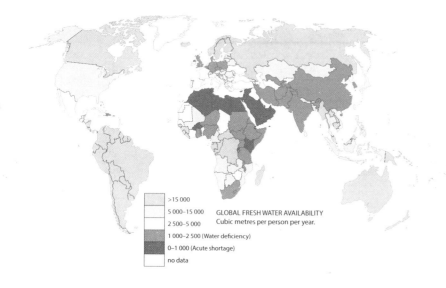

- >15 000
- 5 000–15 000
- 2 500–5 000
- 1 000–2 500 (Water deficiency)
- 0–1 000 (Acute shortage)
- no data

GLOBAL FRESH WATER AVAILABILITY
Cubic metres per person per year.

Activity

Think about all of the things that you do each day that you would struggle to do if you did not have access to clean water. Write a paragraph to explain how your life would be different.

Where there is a **water deficit** people are trying to use water more sustainably. They are trying to reduce the total use whilst meeting people's increasing needs. For example, reducing **irrigation** by 10% could double the amount of water available for **domestic water**.

How can this issue be managed?

Managing water use sustainably can be done on a variety of scales. For example, at an individual level, washing dishes or washing the car by hand uses less water than using a dishwasher or car wash. At a larger scale and in agriculture, **drip-irrigation** uses less water than methods such as **flood irrigation**. However, at a national scale it is often difficult for countries to agree how they will manage the issue. This is because rivers often run through several countries and actions in one place will have knock-on effects elsewhere.

The River Nile passes through 10 countries before reaching Egypt. Egyptian farmers use water from the Nile for irrigation. The country has built dams, such as the Aswan High Dam, to store water from the Nile. This has allowed the growing of crops such as rice and cotton. An agreement has been signed by countries in the Nile basin to promote sustainable use of the river water. However, the population in this area is expected to double to 160 million by 2040. Countries are demanding more water to grow food and generate power for this growing population. This tension has split the countries and could result in conflict.

What case studies do I need?

You need case studies on:

* Water supply in a country or area, e.g. water supply in northern India.

Revision tip

Don't forget that issues such as access to clean water need solutions that operate at different scales from the individual level to the international level.

Activity

Write a list of ways in which you could use water more sustainably. For example, you could turn off the tap while brushing your teeth or take a shower rather than a bath.

Quick test

1. What is an aquifer?
2. What do LEDCs tend to use most of their water for?
3. Give an example of a place that faces a water deficit.
4. What is domestic water?
5. Suggest **one** way in which water can be used more sustainably.

3.7 Environmental risks of economic development

How might economic activities pose a threat to the natural environment?

Economic activities such as agriculture, industry, tourism and power generation have many benefits. However, they also have the potential to harm the natural environment in a number of ways:

- **Soil erosion** – in tropical areas, if deforestation occurs there is nothing to protect the soil from frequent and heavy rain. **Gullies** are formed and the soil is washed away. In drier areas, the wind causes soil to erode. Once vegetation cover has been removed, the soil dries out and is blown away. This soil can be transported huge distances and be washed into rivers. This can reduce water quality, affect wildlife and cause flooding.

- **Desertification** – this is the process where semi-arid environments change to become more desert-like. It is not the spread of deserts, but rather what is happening to the areas next to them. For example, in the Sahel, every year 80 000km² becomes too dry and infertile for farming. Climate change is a major cause of desertification as it increases temperatures and changes seasonal rainfall patterns.

Topic summary

You must be able to:
- Describe how economic activities may pose threats to the natural environment, locally and globally
- Demonstrate the need for sustainable development and management
- Understand the importance of resource conservation.

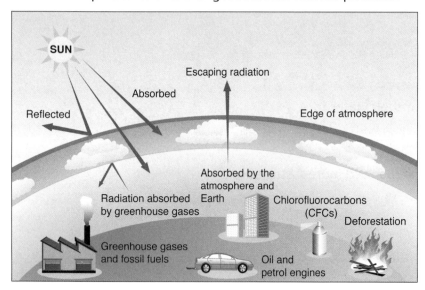

- Enhanced global warming – **greenhouse gases** are produced by burning fossil fuels, deforestation, using agricultural fertilisers and from decaying waste and vegetation. As the world's population grows, so the amount of these gases increases. This traps more heat in the atmosphere and causes temperatures to rise. As a result, sea levels are likely to rise, there are likely to be more frequent and more extreme weather events and coral bleaching could occur.

Melting polar ice caps, ice sheets (in countries like Greenland) and glaciers could cause a sea-level rise of up to 5 metres – a severe problem for the 40 per cent of the world's population who live close to the coast.

Rising sea temperatures will mean the oceans expand, leading to an average worldwide sea-level rise of 40 cm by 2100 – this would mean widespread flooding of low-lying areas.

There could be changes in extreme weather events – e.g. temperatures over 40°C causing wildfires in Australia, more hurricanes in the Caribbean, deadly heatwaves in Europe and floods in Egypt.

Wars may be fought as the global population grows and precious resources such as food and water become scarce in some countries.

THE EFFECTS OF CLIMATE CHANGE

Changes to farming as equatorial areas become too hot and tropical crops are grown further north.

The coral reefs that provide many countries (including Belize, Australia and the Maldives) with a big tourist attraction could die.

Extinction of wildlife that is unable to adapt to the changing climate – some scientists estimate up to one third of all species could become extinct over the next 100 years.

Tropical diseases become more widespread – malaria, for instance, could become common in Europe.

Revision tip

Don't forget that economic activities may have economic or social disadvantages too. It's also worth remembering that economic activities can be beneficial to the natural environment – although the focus of this unit is very much on the risks.

Activity

Carry out some research on a place that has experienced either severe soil erosion or desertification. Create a poster showing what the causes and impacts are.

- Water pollution – water used for industrial cooling can be returned to the river containing **toxic** chemicals. The water is **contaminated** and cannot be used for drinking supplies or irrigation. Similarly, pollutants can enter **groundwater** supplies from factory or power station leaks. These supplies are often pumped to the surface through wells for drinking water.
- Air pollution – smoke from factories and exhaust fumes from cars, lorries and aircraft can cause breathing difficulties for people living nearby. Coal dust produced from mining can have a similar effect. Some chemicals react with water in the atmosphere and fall to the ground as **acid rain**.
- Noise pollution – most people can live safely with noise of up to about 80 **decibels**. This is the equivalent of car traffic or people talking in a crowded room. However, noise which is louder than this, for example, aircraft taking off and landing, can damage people's hearing.
- Visual pollution – some people may consider structures or buildings as **aesthetically damaging**. For example, some people think that wind turbines or new houses can make rural areas visually polluted.

Revision tip

Whilst water, air and noise pollution can all be measured, visual pollution is more **subjective** and different people are affected in different ways. It is therefore sometimes harder to manage than other forms of pollution.

Activity

Is there anywhere near where you live which is polluted? How is it polluted? What has been done to manage the pollution?

Why do we need sustainable management and resource conservation?

The way we use natural resources can be very damaging. However, governments and organisations are looking for alternative ways to be more sustainable. This means using resources in a way that will not leave permanent damage for future generations. Governments and organisations can carry out research into how best to manage resources sustainably and create policy to change behaviours. There are also things that individuals can do to be more sustainable.

- Reduce – reduce the use of natural resources. For example, turn off the heating and wear an extra jumper.
- Reuse – use things you have more often. For example, fill up plastic water bottles at home instead of buying new ones.
- Recycle – make something you own into something else. Over 80% of household waste can be used again. For example, paper, glass, computers and video games can all be recycled.

Bins with paper, plastic, glass, food, metal and electronic waste sorted for recycling

National Parks

National Parks are one way in which we can conserve the natural environment. In most parts of the world National Parks are **wilderness** areas, although in the UK they are found in areas of countryside where people live and work. The main aims of UK National Parks are:

- To maintain and improve natural beauty, wildlife and culture
- To help the public understand and enjoy their special qualities
- To make sure local communities can survive economically and socially.

What case studies do I need?

You need case studies on:

- An area where economic development is taking place, causing the environment to be at risk, e.g. the Aral Sea, Greenland, Borneo or the Great Barrier Reef.

Revision tip

National Parks are one way to conserve the environment on a large scale but they are not the only way. What other methods of large-scale conservation can you think of?

Activity

Research a National Park in your country. Produce a factfile with the main things that you discover.

Quick test

1. What are gullies?
2. Why might noise pollution be hazardous?
3. What is acid rain?
4. Why is recycling sustainable?
5. What are National Parks?

1 The photograph above shows farmers involved in agribusiness. Define what is meant by 'agribusiness'.

.... **[1]**

2 Look at the graph below, which shows the growth of global tourism from 1950–2020.

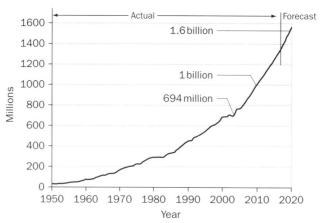

Describe how the total number of tourists has changed over the last 50 years.

....

.... **[2]**

3 Science parks are places where high-tech companies tend to be located. Suggest three characteristics of science parks.

....

,,

.... **[3]**

4 Many parts of the world face a shortage of water. Explain why some areas face these water deficits.

....

....

.... **[3]**

5 Economic activities such as agriculture, industry, tourism and electricity generation all have benefits to the countries in which they take place. However, they can also cause a number of problems. Suggest how economic activity may pose a threat to the natural environment.

....

....

....

....

.... **[4]**

6 The photograph shows a nuclear power station. Describe the benefits of nuclear power.

....

....

....

....

.... **[4]**

7 Transnational Corporations (TNCs) are large, global companies which have their headquarters in an MEDC and usually have a number of factories in LEDCs. Describe and explain the advantages and disadvantages of a TNC you have studied.

....

....

....

....

....

....

....

....

....

....

....

....

....

....   [7]

There are a large number of geographical skills that you have to be familiar with and be able to use and interpret. Don't forget to take a pencil, rubber, ruler, a protractor and a calculator into the exam room with you. It's also useful to have access to a sheet of plain paper for measuring distance or for assisting with cross-sections on the large-scale map.

The Paper 2 Geographical Skills exam is 1 hour and 30 minutes long and the total number of marks available is 60. You could consider allocating about a minute and a half per mark. Roughly, this means that you would be allowing yourself 30 minutes to answer the map question and 50 minutes to answer the other questions. You could then use the remaining 10 minutes to check your answers.

Use the audit below to assess how confident you are with the various geographical skills.

Geographical skill	😀	😀 | 🙁	🙁
Four-figure grid references			
Six-figure grid references			
Compass directions			
Measuring horizontal distances on a map			
Contour reading			
Interpretation of cross-sections			
Translate the scale of a feature by describing its size and shape			
Draw inferences about the physical and human landscape			
Identify and describe basic landscape features			
Describe variations in land use			
Pictograms			
Line graphs			
Bar graphs			
Divided bar graphs			
Histograms			
Kite diagrams			
Flow diagrams			
Wind rose graphs			
Dispersion graphs			
Isoline maps			
Scatter graphs			
Choropleth maps			
Pie graphs			
Triangular graphs			

Geographical skill	😁	😁 | ☹	☹
Radial graphs			
Describe and analyse features from data tables			
Show an understanding of written material			
Describe human and physical landscapes on:			
Photographs			
Aerial photographs			
Satellite images			
GIS			
Describe and annotate field sketches			
Interpreting cartoons			

Now, have a go at the following questions.

1   Study Fig. 5, which shows the structure of the total New Zealand population, and Fig. 6, which shows the structure of the Maori population in 2006. The Maori people form part of the population of New Zealand.

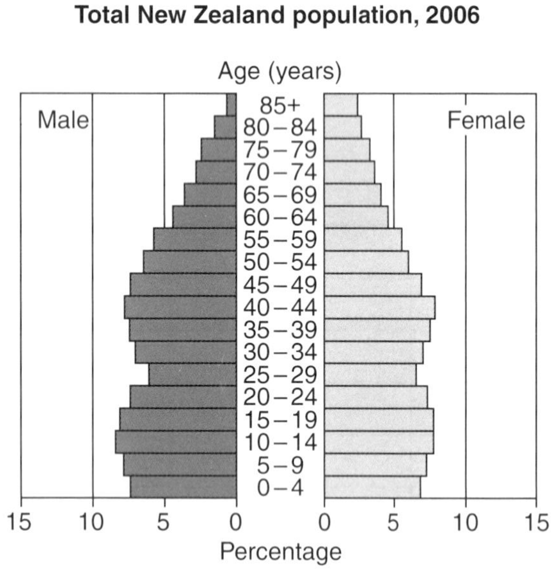

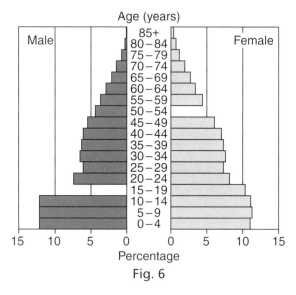

Total New Zealand population, 2006

Maori population, 2006

Fig. 5

Fig. 6

(a)   In 2006, 10% of the male Maori population were aged 15–19 and 5% of the female Maori population were aged 50–54. Complete Fig. 6 by adding this data. [2]

(b)   Complete the following sentences by adding the words **greater** or **less**.

(i)   The percentage of 0–14 year olds in the Maori population is .... than the percentage of 0–14 year olds in the total New Zealand population. [1]

(ii)   The percentage of over 55 year olds in the Maori population is .... than the percentage of over 55 year olds in the total New Zealand population. [1]

(iii) The percentage of over 35–49 year olds in the Maori population is ………… than the percentage of over 35–39 year olds in the total New Zealand population. [1]

(c) In 2006, the Maori population formed 14% of the total New Zealand population.

    (i) Using evidence from Figs 5 and 6 only, suggest how this may change over the next 50 years. [1]

    (ii) Explain your answer to **(c) (i)**. [2] [Total: 8 marks]

(Cambridge IGCSE Geography 0460 Paper 2 Q2 June 2009)

2 Photograph A (below) shows an area of small-scale subsistence agriculture in Asia.

(a) Describe the relief of the area shown in Photograph A. [4]

(b) The natural vegetation of the area is tropical rainforest but the forest has been affected by human activity. Which of the following statements describe the distribution of forest shown in Photograph A? Circle **two** correct statements.

- covering the whole area
- on the highest land
- on the steepest slopes
- in valleys
- completely removed [2]

(c) Soil erosion is a problem in the area shown in Photograph A. What features shown in the photograph may encourage soil erosion? [2] [Total: 8 marks]

(Cambridge IGCSE Geography 0460 Paper 2 Q3 June 2009)

For the Cambridge O Level you will sit Paper 3 (Geographical Investigations), whereas for the Cambridge IGCSE you will have a choice between Paper 3 (Coursework) and Paper 4 which is an alternative examination paper which you sit if you do not do a coursework assignment. The coverage of Paper 3 in the O Level and Paper 4 in the IGCSE are the same and so you can use this section of the revision guide to help you prepare prepare for either exam.

The examination tests:
- Your understanding of the ideas being investigated
- Collection and presentation of data and your analysis of this data
- Your conclusions about the ideas being investigated and suggestions to improve the investigation.

This exam is 1 hour and 30 minutes long. In this time you must answer two questions, each of which is worth 30 marks.

The focus of each question is explained in the introduction, which 'sets the scene'. Each question then includes two hypotheses which describe what is being tested in order to reach a conclusion.

In order to do well in this exam you will need to have experience of fieldwork. This can be carried out around the school site, local area or further afield. You should follow the route to enquiry outlined below.
1. Create a hypothesis or key question.
2. Collect relevant data.
3. Present the data in maps and graphs.
4. Analyse the data you have collected.
5. Draw conclusions about the hypothesis and evaluate the investigation.

It is important that you take care when plotting data because your answers need to be accurate. Look at the two answers (A and B) to each task below and decide which answer is correct.

Task 1: Use the results in the table below to complete Fig. 4, to show the average height of vegetation at points 8, 9, 10 and 11 across the transect at Site A.

sample point	1	2	3	4	5	6	7	8	9	10	11
average height of vegetation (cm)	14	11	7	4	2	0	3	4	5	12	17

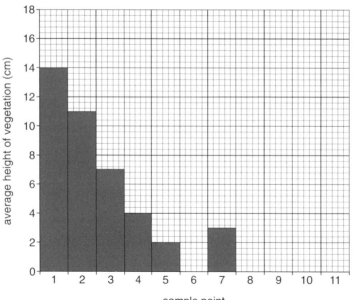

Average height of vegetation at Site A

Fig. 4

Look at the two answers (A and B) and decide which answer is correct.

Answer A

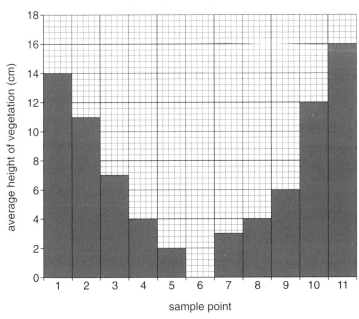

Fig. 4

Answer B

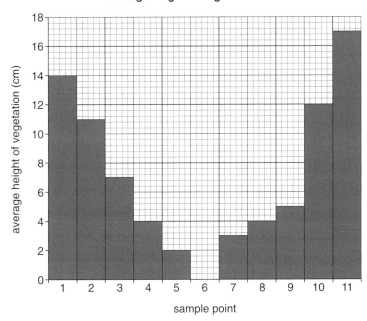

Fig. 4

Task 2: On Fig. 3 shade in the land valued above 60 thousand US$/m².

(Adapted from Cambridge IGCSE Geography 0460 Paper 4 Q1f (ii) June 2007)

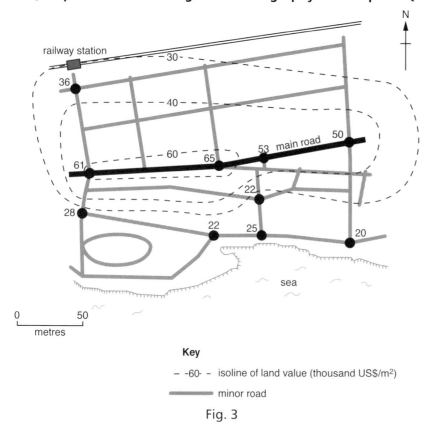

Fig. 3

Look at the two answers (A and B) and decide which answer is correct.

Answer A

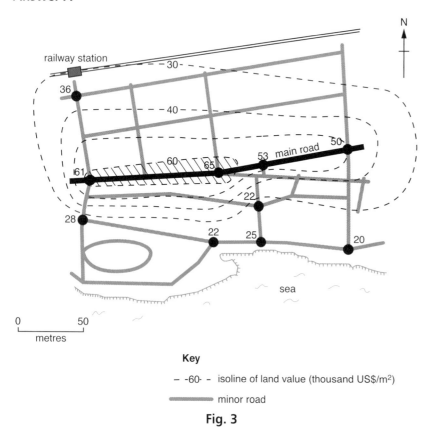

Fig. 3

Answer B

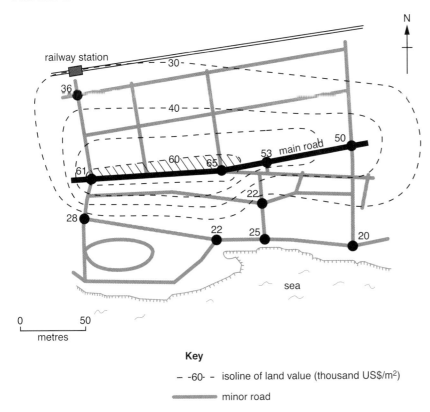

N

railway station

30

36

40

50

53 main road

61 60 65

22

28

22 25 20

sea

0   50
metres

Key

‒ ‒60‒ ‒ isoline of land value (thousand US$/m²)

━━━━ minor road

Fig. 3

Glossary

1.1 Population dynamics

Anti-natalist policies – strategies designed to limit a country's population growth, for example the one-child policy in China.

Birth rate – the average number of live births in a year (for every 1000 people).

Death rate – the average number of deaths in a year (for every 1000 people).

Infant mortality rate – the proportion of children dying at birth or before their first birthday.

LICE – an acronym that stands for: Life expectancy, Infant mortality, Care of the elderly, and Economic.

Migration – the movement of people from one place to another.

Natural change – the birth rate minus the death rate. If the answer is positive, the population is increasing. If the answer is negative then the population is decreasing.

Over-population – when a country does not have the resources to give all of its people an adequate standard of living. For example, Tanzania.

Population explosion – the dramatic rise in world population which took place during the last two centuries.

Pro-natalist policies – strategies designed to stimulate a country's population growth by increasing its fertility rate. For example, France.

Under-population – describes an area where the population is well below that which can be supported by its natural resources. For example, Canada.

1.2 Migration

Destination – the place where migrants are travelling to.

Economic migrants – people who migrate to find work.

Emigrants – people leaving a country.

Forced migration – when people are made to leave a country, for example because of war or natural disaster.

Immigrants – people arriving in a new country.

Origin – the place where migrants come from.

Permanent – where the migrant stays in one place for a long period of time.

Pull factors – factors encouraging people to move to an area.

Push factors – factors encouraging people to move from an area.

Refugees – people who are forced to move, often by war or natural disaster.

Temporary – where a migrant stays in one place for a short period of time before migrating again.

Voluntary – when people choose to leave an area.

1.3 Population structure

Age dependency – the link between the number of adult people who create wealth and the young and elderly population who depend on them for support.

Age/sex pyramid – a diagram that uses horizontal bar graphs to show the age/gender characteristics of a population.

Ageing populations – a country or area with a large proportion of elderly dependents.

Elderly dependents – elderly people who are retired and dependent upon the working population.

Materialistic – as wealth increases people become more concerned with material possessions.

Pensions – money paid to people after retirement.

Population structure – the way a population is composed of different age/gender groups.

Youthful dependents – young people who do not work and who are dependent upon the working population.

Youthful populations – a country or area with a large proportion of youthful dependents.

1.4 Population density and distribution

Cash crops – a crop grown to sell for profit.

Densely populated – describes an area where a lot of people live (often a large number of people per km²).

Desertification – the process in which semi-arid environments change to become more desert-like.

Extreme environments – environments which have very high or very low temperatures, very high or very low amounts of precipitation or which are mountainous can be described as extreme. It is difficult for people to live there.

Famine – the long-term situation where people do not have enough food to eat.

Population density – the number of people living in a 1km² area.

Population distribution – the way in which a population is spread out over an area.

Soil degradation – the process of soil becoming less fertile.

Sparsely populated – describes an area (usually in km²) where few people live.

Uninhabited – describes an area where very few, if any, people live.

1.5 Settlements and service provision

Catchment area – the area served by the shops and other facilities in an urban settlement. Also known as the sphere of influence.

Dispersed – describes the way that isolated farms or houses are scattered over an area, far apart from each other.

Function – the way that a place provides employment and services for the people who live in or visit it.

Linear – describes long, narrow-shaped settlements, often built along a road.

Nucleated – describes a 'compact' settlement, about the same width in most directions, built around a central point such as a road junction or river crossing.

Settlement – any place, ranging in size from a hamlet to a mega-city, where people live and work.

Settlement hierarchy – arranging settlements according to their population size.

Site – a small area of land on which the first part of a settlement was built.

Situation – the location of a settlement in relation to the wider area around it.

Sphere of influence – the area served by the shops and other facilities in an urban settlement. Also known as the catchment area.

Threshold population – the minimum number of people needed to support a facility or service.

1.6 Urban settlements

Brownfield site – a site of derelict or disused land in an urban area.

Central Business District (CBD) – the central and most accessible zone within a large settlement, which has many offices, large shops and public buildings.

Commute – to travel from home to a place of work on a regular basis.

Congestion charges – a charge imposed on vehicles entering a central urban zone to reduce its traffic congestion and level of air pollution.

Greenfield site – a site which has not previously been built on.

Infrastructure – the network of services needed for industry and services to run efficiently – transport, communications, energy supplies, water and sanitation systems.

Land use zones – urban areas which have different functions and characteristics to other nearby areas.

Pollution – what happens when the environment is harmed; its four types are air, noise, visual and water pollution.

Quality of life – an indicator of how people assess their lifestyle based on criteria which include GDP per capita and life expectancy.

Rapid transit systems – a transport network, such as an underground railway, designed to carry large numbers of people within an urban area.

Redevelopment – raising housing standards by demolishing and then replacing existing accommodation.

Rural–urban fringe – this zone is found at the edge of an urban area where the city meets the rural area.

Smog – air pollution due to a combination of smoke and fog.

Squatter settlement – an area of poorly built, low-cost housing on land not owned or rented by its inhabitants.

Urbanisation – an increase in the percentage of people living in urban areas.

Urban sprawl – the outward growth of a built-up area into nearby rural areas.

1.7 Urbanisation

Informal labour – employment that has no set hours or employment benefits; its low, irregular earnings are not taxed.

Over-population – when a country does not have the resources to give all its people an adequate standard of living.

Pull factors – factors encouraging people to move to an area.

Push factors – factors encouraging people to move from an area.

Self-help scheme – a way of raising housing conditions for squatters by providing them with some land and the basic materials needed to build and improve their own homes.

Site and service scheme – a way of raising housing conditions for squatters by providing them with basic accommodation and facilities such as water supply and sewage disposal.

Sustainable – something that can be used on a long-term basis with very little effect on the environment.

Urbanisation – an increase in the percentage of people living in urban areas.

2.1 Earthquakes and volcanoes

Conservative – an area where two plates slide past each other.

Constructive – where two plates move away from each other.

Convection currents – the mantle is heated by the core. This heat is transferred by convection currents. As the bottom of the mantle is heated it rises to the top, it then cools and sinks again to the bottom.

Crater – a bowl-shaped depression with steep sides formed by a volcanic eruption.

Crust – the top layer of the Earth; it can be either continental (with land on top) or oceanic.

Destructive – where an oceanic plate slides under a continental plate.

Epicentre – point on the Earth's surface directly above where an earthquake occurs.

Focus – the point within the Earth's crust where an earthquake originates.

Inner core – the centre of the Earth and the hottest part of the planet. It is a solid ball of iron and nickel and with temperatures of around 5400°C.

Lava – hot, molten rock which is erupted from a volcano.

Magma – liquid rock found beneath the Earth's crust.

Magma chamber – a reservoir of magma in the Earth's crust found beneath a volcano.

Main vent – this is the main way in which magma escapes from a volcano.

Mantle – the part of the Earth between the core and the crust.

Outer core – a liquid layer composed of iron and nickel which lies outside the inner core.

Plate boundaries – also known as plate margins, these are the edges of tectonic plates.

Pyroclastic flow – flow of materials such as ash ejected during a volcanic eruption.

Seismic waves – a wave which passes through the Earth as a result of an earthquake.

Shield volcano – a large, gently-sloping volcano formed from thin, runny lava and frequent gentle eruptions.

Strato volcano – also known as a composite volcano, made up of alternate ash and lava layers. Strato volcanoes often have very violent eruptions.

Subduction zone – where oceanic crust is forced under continental crust.

Tectonic plates – the Earth's crust, which is divided into large pieces, called plates.

2.2 Rivers

Attrition – when boulders and large stones carried by the river bash into each other and break up into smaller pieces.

Corrasion – when sand and stones carried by the river rub against the bank and bed and knock off other particles.

Corrosion – when acids in the river dissolve the rocks that make up the bank and bed.

Delta – the area where silt is deposited as a river enters the sea (or a lake).

Drainage basin – the area of land drained by a river.

Evaporation – the change of state from water droplets (liquid) to water vapour (gas) caused by heating.

Groundwater flow – underground water supplies.

Hydraulic action – when the force of the water knocks particles off the sides (banks) and the bed of the river.

Infiltration – the downward movement of water into the soil.

Interception – precipitation falling into the drainage basin can be intercepted by the leaves of trees and by ground vegetation.

Load – all the material carried by a river.

Overland flow – some of the precipitation reaching the ground flows over the surface until it reaches a river channel.

Throughflow – infiltrating water continues down slopes through the soil towards the river channel.

2.3 Coasts

Backwash – the movement of water back down a beach towards the sea.

Bay – indentation in the coastline where wave action erodes softer rock.

Beach nourishment – this is the process of putting sand from elsewhere onto an eroding beach to create a new beach or widen an existing beach. It is an example of soft engineering.

Constructive waves – low-frequency waves of low height, with a strong swash but weak backwash; the waves therefore build up material on a beach.

Coral bleaching – the whitening of coral due to expulsion or death of their symbiotic algae, usually as a result of changing environmental conditions.

Destructive waves – high-frequency, steep waves which have little swash and so move little material up a beach; however, they have strong backwash and so drag material down the beach into the sea.

Dune stabilisation – this involves planting vegetation on sand dunes so that the roots bind the sand together. This makes the dunes more resistant to erosion. This is a soft engineering technique.

Fetch – the distance across open sea or ocean over which the wind blows to create waves; the longer the fetch, the greater the possibility of large waves.

Groyne – a low wall, usually made from wood, which sticks out into the sea. It prevents longshore drift and is an example of hard engineering.

Headland – point along a coast where harder rock juts out into the sea.

Lagoon – area of salt water separated from the sea by a bar or reef.

Longshore drift – the transportation of material along a coastline.

Overfishing – when the stock of fish in the sea is depleted from too much fishing.

Polyps – tiny marine organisms that live in and build coral reefs.

Sea wall – a form of hard engineering where a wall is built to protect the coastline from erosion.

Soft engineering – this involves working with nature to try to manage the coastline.

Swash – the body of water rushing up a beach after a wave has broken.

2.4 Weather

Air pressure – the pressure at a point on the Earth's surface due to the weight of the air above that place.

Anemometer – an instrument used to measure wind speed.

Barometer – an instrument used to measure atmospheric (air) pressure.

Beaufort Scale – a scale that measures wind intensity based on observing conditions at sea.

Cloud – a cloud is a visible mass of water droplets or ice crystals suspended in the atmosphere. Clouds form when air is unable to absorb any more water – it becomes saturated. This usually happens when air cools because cooler air cannot hold as much water vapour as warm air. Therefore, condensation occurs and clouds form.

Hygrometer – an instrument used to measure relative humidity.

Maximum-minimum thermometer – an instrument used to measure both the highest and lowest temperatures within a 24-hour period.

Meteorologist – a person who is able to forecast (predict) future weather by referring to previous, similar atmospheric conditions.

Relative humidity – this describes the amount of water vapour in the air. Air contains water vapour due to evaporation and transpiration. How much water vapour it holds depends upon its temperature. Warm air can hold more water vapour than cold air.

Stevenson Screen – a wooden box on legs used to house weather recording equipment.

Sunshine recorder – records the hours of sunshine by using a glass sphere that focuses the Sun's rays to burn a track on a paper strip.

Weather – short-term, changeable, atmospheric conditions which include air temperature, cloud cover, precipitation and wind direction and speed.

Wind vane – an instrument used to indicate the direction of the wind.

2.5 Climate and natural vegetation

Altitude – the height above sea level.

Arid – dry. There is a lack of available water meaning plant growth is difficult. Desert environments are arid.

Biodiversity – the range of species within an ecosystem.

Convectional rainfall – precipitation formed by rising currents of warm, moist air.

Deforestation – the removal of tree cover in an area for farming or other activities.

Diurnal temperature range – the difference between the highest and the lowest temperature in a 24-hour period.

Greenhouse gases – gases in the atmosphere that act like a blanket and reflect heat.

Habitats – the 'home' of a plant or animal.

Insolation – solar energy which reaches the Earth's surface.

Latitude – the distance north or south of the Equator.

Leaching – the removal of nutrients and other minerals from the soil as rainwater washes the minerals downwards through the soil.

Leaf litter – dead plant material that falls to the ground, releasing nutrients as it decomposes.

Nomadic – a lifestyle that involves moving around with no permanent home.

Pastoral farmers – a type of farming which rears livestock rather than grows crops.

Radiation – the transfer of heat energy in the form of electromagnetic waves.

3.1 Development

Brazil, Russia, India and China (BRICs) – an acronym which describes four countries that are at a similar stage of development and are newly emerging economies.

Exports – the goods that a country sells overseas.

Human Development Index (HDI) – a system of ranking countries based on their GDP (gross domestic product) per capita and rates of adult literacy and life expectancy.

Indicators of development – show a country's progress economically as well as in other social aspects.

Informal sector – employment which has no set hours or employment benefits; its low, irregular earnings are not taxed.

Least Developed Countries (LDCs) – the poorest countries in the world.

Less Economically Developed Countries (LEDCs) – one of the poorer countries in the world.

More Economically Developed Countries (MEDCs) – one of the wealthier, more industrialised countries in the world.

Multiplier effect – the additional economic effects experienced when money is spread throughout a community.

Newly Industrialising Countries (NICs) – countries that are developing their economies through rapid expansion of secondary industries.

Raw materials – the things that are used to make something else.

Sustainable Development Goals (SDGs) – a set of targets produced by the United Nations in 2015 designed to reduce global poverty and improve the quality of life in many countries.

Transnational Corporation (TNC) – a very large business operating in many countries.

3.2 Food production

Agribusinesses – large companies involved in agriculture, often with many different farms.

Arable – farms which grow crops.

Calories – the unit used to measure the energy we get from food.

Commercial – farms which produce food for sale to make a profit.

Extensive – farms which have smaller inputs but usually use more land.

Inputs – the things that go into a system.

Intensive – farms that use large amounts of money, machines and technology or workers.

Irrigation – artificially diverting water to fields to grow crops.

Labour – workers.

Malnutrition – the condition that results when the body does not get enough vitamins, minerals or other nutrients.

Mixed – farms that grow crops and rear animals.

Nomad – person whose lifestyle involves moving around with no permanent home.

Outputs – the things that come out of a system.

Pastoral – farms which rear animals.

Plantations – a large area of land which has been deforested and the trees replaced with a single type of crop.

Processes – activities carried out to turn inputs into outputs.

Sedentary – settled and permanent.

Shifting cultivation – a type of rainforest agriculture which involves moving on from one plot of cultivated land to another.

Subsistence – a crop grown to feed a farmer and his/her family.

3.3 Industry

Assembly – putting things together, e.g. assembling a car from its constituent parts.

Entrepreneurship – someone who sets up a business and takes the financial risk of doing so.

Feedbacks – outputs from an industrial system that can be used as inputs, e.g. waste for recycling or as an energy source or profit which can be reinvested as capital.

Footloose – an industry that is not tied to specific location factors.

Graduates – people with a university degree.

Grants – amounts of money which can be given to businesses for specific purposes.

High-tech industries – industries concerned with the application of technology. For example, computers, telecommunications and computer gaming.

Manufacturing – to make something on a large scale using machinery.

Processing – the production of, for example, food, drink, chemicals or textiles.

Raw materials – the things that are used to make something else.

Secondary sector – industries that make products from raw materials.

Science parks – these are centres where many high-tech industries are found.

Site – a small area of land on which an industry is located.

3.4 Tourism

Adventure tourism – holidays involving physical activities such as white water rafting or sky diving.

Disposable income – income remaining after taxes, leaving money which can be spent on things such as holidays or consumer goods.

Ecotourism – a type of tourism which focuses on tourists experiencing the natural environment.

Infrastructure – the network of services needed for industry and services to run efficiently – transport, communications, energy supplies, water and sanitation systems.

Management – where people change a natural environment in order to try to make it sustainable.

Multiplier effect – the additional economic effects experienced when money is spread throughout a community.

Package holiday – a holiday organised by a travel agent, with arrangements for transport, accommodation, etc. made at an inclusive price.

Seasonal employment – jobs which are only available for part of the year. They are usually temporary, short-term and poorly paid.

Sustainable – something that can be done on a long-term basis with very little effect on the environment.

Tourism – the industry associated with people going on holiday; can be either national or international.

Westernised – to be influenced by the cultures of the west, e.g. USA and Europe.

Wilderness tours – tours to places that are uncultivated, uninhabited and inhospitable regions.

3.5 Energy

Consumption – the amount of energy which is used by a country or region.

Energy mix – the combination of different energy sources that a country or region uses.

Fossil fuels – energy resources such as oil, gas and coal.

Non-renewable energy – energy which is used once and cannot be used again.

Nuclear reactor – the part of a nuclear power plant where a controlled nuclear reaction takes place in order to release energy.

Renewable energy – sources of power that can be used over and over again.

Turbines – a rotor which is powered by water or wind to provide a continuous supply of energy.

3.6 Water

Aquifer – a layer of porous rock which stores underground water.

Basic human needs – things like food, water and shelter that people need in order to survive.

Desalinisation – removing salt from sea water so that it can be used as drinking water.

Domestic water – water used in the home.

Drip-irrigation – where small amounts of water are used to target the roots of crops.

Flood irrigation – where large amounts of water are used to grow crops. The area is simply flooded with water.

Irrigation – artificially diverting water to fields to grow crops.

Virtual water – water which is used in the process of manufacturing things but which is 'invisible' i.e. part of the process but not seen in the finished item.

Water deficit – a situation where available water supplies do not meet all the needs of local people.

3.7 Environmental risks of economic development

Acid rain – rainfall that damages the environment because it has been made acidic by pollution in the atmosphere.

Aesthetically damaging – where something does not look nice.

Contamination – when pollutants are added to the natural environment.

Decibels – how sound volume is measured.

Desertification – the process in which semi-arid environments change to become more desert-like.

Greenhouse gases – gases in the atmosphere that act like a blanket and reflect heat.

Groundwater – underground water supplies.

Gullies – channels in the soil caused by water erosion.

Soil erosion – where soil is washed away by water or blown away by the wind.

Subjective – a matter of opinion.

Toxic – poisonous.

Wilderness – an uncultivated, uninhabited and inhospitable region.

Population and settlement

1.1 Population dynamics

1. Life expectancy, infant mortality, care of the elderly, economic
2. Over-population – when a country does not have the resources to give all of its people an adequate standard of living, e.g. Tanzania.
   Under-population – describes an area where the population is well below that which can be supported by its natural resources, e.g. Canada.
3. **Any two from:** human conflict; infant mortality rate; good sanitation; nutrition; clean water; natural disasters.
4. **Any two from:** emancipation of women; education; urbanisation; average age of marriage; culture; religion.
5. A population policy is where the government of a country imposes laws to try to control their population growth rate. An example of an anti-natalist population policy is China's one-child policy.

1.2 Migration

1. An immigrant is someone who is moving into a country. An emigrant is someone who is leaving a country.
2. A refugee is a person who is forced to move, often because of war or natural disaster.
3. **Any two from:** migrants may be employed doing menial jobs; the destination country may be able to gain skilled labour cheaply; a multi-ethnic society may increase understanding and tolerance of other cultures.
4. **Any two from:** migrants are usually healthy young men who would be capable of doing useful work at home; a gender imbalance is created with more women than men being left behind; many emigrants are educated and the population left behind are less able to build a better country; the young and the elderly are left behind, putting pressure on both the education and healthcare systems.
5. The answer will depend upon the specific case study chosen. However, the answer should include the name of the country of origin and destination as well as **two** impacts on the migrants themselves.

1.3 Population structure

1. At stage 3, birth rates are reducing and death rates are low.
2.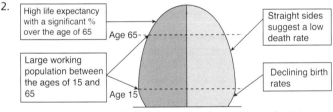

| High life expectancy with a significant % over the age of 65 | | Straight sides suggest a low death rate |
| Large working population between the ages of 15 and 65 | | Declining birth rates |

3. Age dependency is the link between the number of adult people who create wealth and the young and elderly population who depend on them for support.
4. **Any from the following:** a large and cheap workforce is created; a large population can provide a growing market, which is attractive to exporting countries; a significant tax base may be created by the large numbers of working people.
5. **Any from the following:** increase in elderly people puts a strain on healthcare services; many countries face a pensions crisis where there is not enough money to cover the pensions of an increasingly elderly population; fewer people of working age can lead to a shrinking economy and a decrease in the amount of tax being paid.

1.4 Population density and distribution

1. Population distribution is the way in which people are spread out over the Earth's land surfaces. Population density is a measure, which calculates how many people live in an area (usually 1km²) and indicates whether an area is sparsely or densely populated.
2. This may depend on the case study that has been selected for study, but expect to see references to extreme climates, mountainous terrain, soil degradation and desertification.
3. This may depend on the case study that has been selected for study, but expect to see references to flat land, fertile soil, fossil fuels, fishing, fresh water, temperate climate, mineral deposits and coastal locations.
4. The process in which semi-arid environments change to become more desert-like.
5. The process of soil becoming less fertile.

1.5 Settlements and service provision

1.
   House —— Road

2. The site of a settlement is the area of land on which it is first built whereas the situation is the settlement's location in relation to the wider area around it.
3. This is the minimum number of people needed to support a facility or service. For example, the threshold population for a university is 100 000 people.
4. 500 people
5. Any settlement with a population of over 500 people should be able to support a primary school. Therefore, most villages, towns, cities, conurbations and mega-cities would be able to support at least one primary school.

1.6 Urban Settlements

1. Central Business District (CBD); industrial zone; residential zone; rural urban fringe
2. **Any two from:** dominated by housing; cheaper land; large plots with gardens and garages; population have a high quality of life and commute to work (more economically developed countries); dominated by squatter settlements; population have a low quality of life; poor housing (less economically developed countries).
3. This is what happens when the environment is harmed. There are four main types of pollution found in urban areas: air, noise, visual and water pollution.
4. **Any two from:** inequalities; traffic congestion; housing issues; conflicts over land use.
5. Answer will depend upon the problems identified.

1.7 Urbanisation

1. There has been an increase in rural to urban migration in less economically developed countries.
2. **Any two from:** poor building construction; disease; risk of fire; unsuitable location; crime; lack of healthcare; lack of education; informal labour.
3. Some common examples are: Roçinha, Rio de Janeiro; Dharavi, Mumbai; 10th of Ramadan City, Cairo and Kibera, Nairobi.
4. These schemes are similar in that they both involve the authorities however, site and service schemes are on a larger scale. They also have the main infrastructure provided, e.g. water, electricity and sanitation, whereas self-help schemes are unlikely to have this.

Answers

5. A range of answers could be accepted here. Examples from the Curitiba case study include: improving public transport; providing urban parks; planting trees; recycling; using renewable energy.

Exam-style questions – Theme 1

1. The east of Kenya has the lowest population density [1].
2. Migration is the movement of people from one place to another [1]. An example of migration is international migration from Senegal to Europe [1].
3. Settlements typically develop into three types: linear, dispersed, and nucleated. Linear settlements are long and narrow and tend to grow up along a road or river [1]. Dispersed settlements are individual farms and houses that are scattered over a rural area [1]. Nucleated settlements are clustered around a road junction or bridge [1].

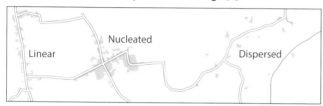

4. Population growth = birth rate – death rate +/- migration so for Kenya 26.4 – 6.89 – 0.22 = 19.29 so the population is growing by 19.29 people per 1000 per year. [Need to show working and have the correct units for the full 3 marks.]
5. [2 marks available for a correctly sketched pyramid, 2 marks available for clear annotations.]

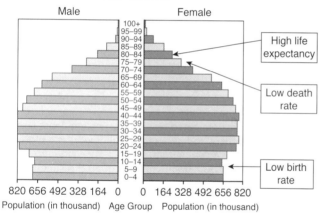

6. Problems could include: traffic congestion, squatter settlements, air and noise pollution.
   Solutions could include: pedestrianised areas, congestion charges, self-help schemes, site and service schemes, public transport. [Five points need to be made to gain 5 marks. A balance is needed between problems and solutions. A maximum of 4 marks will be awarded if only one aspect is tackled.]
7. Levels of marking applied to urban example:
   Level 1 (1–3 marks): Statements including limited detail that suggest characteristics of an urban area which is rapidly growing and ways in which the urban area could be made more sustainable, e.g. disease, poor building construction, risk of fire, crime, lack of services, site and service schemes, self-help schemes.

Level 2 (4–6 marks): More developed statements that explain reasons for characteristics and ways in which the urban area could be made more sustainable, e.g. disease because sanitation is poor and people live in close proximity to each other / crime because many people are unemployed or have jobs in the informal sector, etc.
Level 3 (7 marks): Three or more developed statements and one named example with at least one piece of place-specific detail, e.g. in Dharavi in Mumbai most of the dwellings have no water or electricity so people have to rely on others to provide what they need, etc.

The natural environment

2.1 Earthquakes and volcanoes

1.

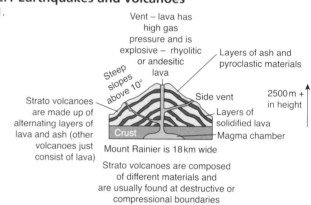

2. Any two from: the crust is the outer layer of the Earth; it is made up of dense but thin oceanic crust; it has thick but less dense continental crust; it is solid.
3. A constructive plate boundary is where two plates are moving apart from each other and so crust is created.
4. Any one from: fertile soils, which are good for farming; tourism; geothermal energy production; mineral mining.
5. Economic impacts are those which affect either money or jobs. So, an economic impact could be any one from: the damage caused needs to be repaired, which can cost a huge amount of money; agriculture or industry is damaged, which has knock-on effects on the economy.

2.2 Rivers

1. Overland flow is where some of the precipitation reaching the ground flows over the surface until it reaches a river channel.
2. Any two from: hydraulic action; corrasion; corrosion; attrition.
3. Either meanders or ox-bow lakes. Waterfalls would also be acceptable.
4. False. A levee is a naturally formed ridge on the edge of a river channel.
5. Opportunities could include any one from: tourism; industry; agriculture; power generation.
   Hazards could include any one from: flooding and associated damage; loss of life, etc.

2.3 Coasts

1.

The formation of headlands and bays

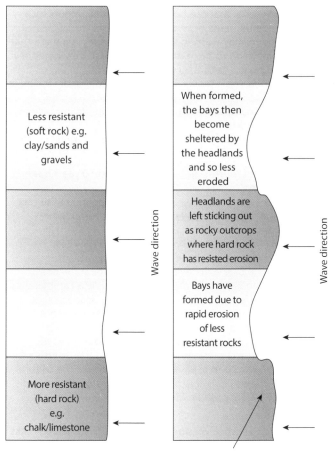

Less resistant (soft rock) e.g. clay/sands and gravels

More resistant (hard rock) e.g. chalk/limestone

Wave direction

When formed, the bays then become sheltered by the headlands and so less eroded

Headlands are left sticking out as rocky outcrops where hard rock has resisted erosion

Bays have formed due to rapid erosion of less resistant rocks

Wave direction

Once formed the headland is then left more vulnerable to erosion and the wave's energy is concentrated here

2. A stump is formed when the coastline is being eroded. Firstly, destructive waves attack the headland and small cracks appear. These cracks get larger as erosion continues, firstly forming a wave-cut notch, a cave and then an arch. After a period of stormy weather, the roof of the arch may collapse leaving a stack, which is a pillar of rock separated from the coastline. A stump is a stack that has been eroded further so that it is covered by the sea during high tide.

3. Barrier reefs and atolls look similar but atolls are associated with volcanic islands and are only found around submerged oceanic islands.

4. The whitening of coral due to expulsion or death of their symbiotic algae, usually as a result of changing environmental conditions.

5. Possible examples include sea walls, groynes or any other strategies which involve building man-made structures.

2.4 Weather

1. A wooden box on legs used to house weather-recording equipment.

2. A sunshine recorder records the hours of sunshine by using a glass sphere that focuses the Sun's rays to burn a track on a paper strip. The total length of the burn shows how long the Sun has been shining so the longer the burn, the longer the hours of sunshine.

3. A hygrometer measures relative humidity, which describes the amount of water vapour in the air.

4. An anemometer. It measures wind speed in km/hr whilst the Beaufort Scale estimates wind speed.

5. Cloud cover is measured in oktas or eighths

2.5 Climate and natural vegetation

1. The climate in the tropical rainforest consists of warm temperatures (usually 28°C all year round) and high rainfall (often more than 2000mm per year). There are no distinct seasons and rainfall is convectional.

2. Leaching is the removal of nutrients and other minerals from the soil as rainwater washes the minerals downwards through the soil.

3. Deserts are very hot during the day, temperatures are usually around 40°C or 50°C. However, temperatures at night are very low and often fall below freezing. Deserts are very arid, meaning that they are very dry, there is little rainfall.

4. Whilst the high temperatures, long hours of daylight and unbroken sunshine are all good for plant growth, rainfall is light, unreliable and unpredictable. Soils are baked hard, making infiltration difficult. This, combined with high evapotranspiration rates, means there is little water available for plant growth.

5. Any of the following would be appropriate answers:
Dormancy – drought-resistant seeds lie dormant until a period of rainfall. These plants can complete their lifecycle within a few weeks.
Water retention – some plants store water in stems, trunks or leaves.
Tolerance of saline conditions – desert soils are salty because evaporation draws salt upwards towards the surface. Salt is toxic to plants but some have developed salt tolerance.

Exam-style questions – Theme 2

1. Coral bleaching is the whitening of coral due to the expulsion or death of their symbiotic algae, usually as a result of changing environmental conditions **[1]**.

2. Day 1 – 18°C; Day 2 – 13°C; Day 3 – 11°C; Day 4 – 11°C **[1 mark for every two correct answers.]**

3. At a conservative plate boundary plates are rubbing against each other **[1]**, the plates are either moving in different directions or in the same direction but at different speeds. Friction builds up as the plates move **[1]** and sometimes one of the plates jars against the other. When this pressure is released an earthquake occurs **[1]**.

4.

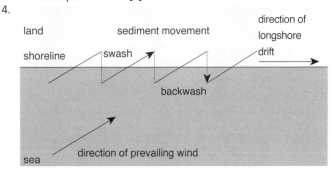

land   sediment movement   direction of longshore
shoreline   swash   drift
                                      backwash
sea   direction of prevailing wind

[1 mark for sketch, 1 mark for prevailing wind, 1 mark for swash and backwash labelled correctly.]

5. Hot deserts have minimal amounts of rainfall, less than 250mm per year **[1]**. Hot deserts also have temperatures which are high all year round and often exceed 40°C **[1]**. However, there is often a huge diurnal range, with temperatures reaching 40°C during the day and –40°C at night **[1]**. There is often a seasonal pattern to the desert climate with one cooler and wetter season and another hot, arid season **[1]**.

6. Named example, e.g. Montserrat – 23 deaths and over 100 injured, mass evacuation of population, which was 12 000 (1995), reduced to 1500 (2001), serious skills shortage due to migration, airport and port closed, economy based on farming, fishing and tourism destroyed, housing shortage led to 70% rent increase. **[1 mark for each point made, named example needed otherwise marks limited to a maximum of 4, also need to cover both social and economic impacts otherwise marks limited to a maximum of 4.]**

7. Levels of marking applied to river management example:
   Level 1 (1–3 marks): Statements including limited detail which suggest how a river could be managed, e.g. dams and reservoirs, afforestation, levees, flood forecasting, diversionary spillways.
   Level 2 (4–6 marks): More developed statements which explain how a river could be managed, e.g. afforestation involves the planting of trees so rainfall is intercepted, which delays runoff into the river, etc.
   Level 3 (7 marks): Three or more developed statements and one named example with at least one piece of place-specific detail, e.g. On the Mississippi River diversionary spillways have been constructed. These are overflow channels which store excess water in times of flood and release it after the floods have passed, etc.

Economic development

3.1 Development
1. The HDI is the Human Development Index. It describes social and economic well-being. It uses adult literacy, life expectancy and GDP per capita to give a score between 0 and 1. Countries with a score close to 0 have low human development whilst countries with a score close to 1 have high human development.
2. Any service industry. For example, healthcare, education or any job involved in tourism.
3. Employment which has no set hours or employment benefits; its low, irregular earnings are not taxed.
4. Most TNCs are based in MEDCs such as the USA and the UK. Unilever, McDonalds and Apple are all examples of TNCs.
5. **Any one from:** TNCs invest in countries by developing the local infrastructure and creating jobs; TNCs usually pay higher wages than local businesses. Employees therefore have a higher quality of life; TNCs bring wealth to the local economy, which has a positive multiplier effect; local people learn new skills and use new technologies; most manufactured products are exported, which benefits the economy; TNCs pay tax to the host country – this can be invested in health, education and infrastructure.

3.2 Food production
1. Farms that use large amounts of money, machines and technology or workers.
2. Anything which is a result of farming. For example, wool, milk, eggs, maize, barley, waste or money.
3. A disease that people may get if there are food shortages and they are malnourished. Symptoms include the stomach swelling, skin peeling and hair turning orange.
4. Crops that are genetically modified. This means the altering of crops to withstand pests and diseases. GM crops have the potential to increase the amount of food that can be grown. However, some countries have banned GM crops until they are proved safe for people and the environment.
5. Giving aid makes people dependent on the food packages being handed out. Aid does not provide a long-term solution to food shortages.

3.3 Industry
1. These are things that can be returned as inputs so, for example, waste that can be recycled or used as a source of energy or capital that can be used to buy other inputs such as electrical components.
2. Entrepreneurs may want to site their industries in a place that means something to them, for example the place where they live or the place where they were born.
3. Raw materials and a power supply were traditionally the most important factors affecting industrial location.
4. An industry that is not tied to specific location factors, it can be located anywhere.
5. A centre where many high-tech industries are found.

3.4 Tourism
1. The number of tourists globally over the last 50 years has increased dramatically.
2. A variety may be acceptable. Answers mentioned in the text include: package holidays, adventure tourism, wilderness holidays and ecotourism.
3. Income remaining after taxes, leaving money that can be spent on things such as holidays.
4. People will be paid and have a job during the summer but not have a job or be paid during the winter. This can make it difficult to budget and survive during the closed season. If people do not have a steady income, then they may not spend money in the local area.
5. A variety may be acceptable and it is likely that they will be linked to the case study. An example would be: posters could be produced to explain to tourists some of the cultural customs. This is sustainable because it means that it reduces conflict between locals and tourists.

3.5 Energy
1. Countries with active volcanoes use energy from heated rocks and magma. In Iceland geothermal power stations produce a quarter of the country's electricity and 90% of heating and hot water.
2. Oil (36%)
3. Energy consumption has increased dramatically as population has increased and our use of gadgets which use energy has also increased.
4. The technology to develop nuclear power is expensive and so it is not really available to poorer countries. Also there have been concerns over the safety of nuclear power because of high-profile nuclear power disasters such as Chernobyl.
5. HEP can slow river flow rates. This can lead to a build-up of sediment, which can affect conditions for wildlife.

3.6 Water
1. An aquifer is a layer of porous rock which stores underground water.
2. Whilst there are exceptions, LEDCs tend to use most of their water for agriculture or in the home.
3. Any country that is shaded in the 30–69% area of the map is likely to face a water deficit as a large proportion of the population do not have access to clean water. This includes countries in sub-Saharan Africa and countries such as Madagascar and Kenya.
4. Domestic water is water that is used in the home.
5. There are a number of answers to this question. Suggestions should demonstrate that they are sustainable and conserve water, for example drip-irrigation, which uses a small amount of water and targets the roots of the plants.

3.7 Environmental risks of economic development

1. Gullies are channels in the soil caused by water erosion.
2. If noise pollution is sustained over a long time or very loud, it can cause hearing loss.
3. Acid rain is rainfall that damages the environment because it has been made acidic by pollution in the atmosphere.
4. Recycling is sustainable because it turns waste into something that can be used again. This means that raw materials are not used up.
5. National Parks are usually areas of wilderness that are protected.

Exam-style questions – Theme 3

1. Agribusiness is where large companies are involved in agriculture, often owning many different farms **[1]**.
2. The number of tourists has grown dramatically over the last 50 years **[1]**.
   For example, in 1960 there were approximately 50 million tourists – this had risen to 1 billion tourists by 2010 **[1]**.
3. Science parks usually have close links with universities, large amounts of capital to buy land and build specialist laboratories and super-fast internet connections. **[1 mark per point, up to a maximum of 3 marks.]**
4. Some areas of the world such as Somalia in eastern Africa have a water deficit **[1]**. This is partly because this is an arid area and precipitation levels are low **[1]** but also because storing water and moving it to areas that need it is expensive **[1]**.
5. There are many ways in which economic activity may pose a threat to natural environments. For example, because the vast ice sheet which covers Greenland has started to melt it is easier to access the huge wealth of mineral deposits that lie beneath the ice sheet. The price of various metals has also increased, making it even more attractive for companies to open mines in Greenland. The impact of agriculture, transport and deforestation on natural environments could also be discussed. **[1 mark per point, up to a maximum of 4 marks.]**
6. The benefits of nuclear power are that the amount of raw material needed is very small compared with other fuels such as coal or oil. Also, nuclear power contributes relatively little to acid rain, global warming and climate change. Many people have concerns over the safety of nuclear power, yet there is a lot of research being carried out to address these concerns and many measures have been taken to make sure it is as safe as possible. Importantly, nuclear power allows countries to reduce their dependence on fossil fuels and to cut imports of these resources.
   [1 mark per point, up to a maximum of 4 marks.]
7. Levels of marking applied to an example of a TNC:
   Level 1 (1–3 marks): Statements including limited detail which suggest the advantages and disadvantages of TNCs, e.g. TNCs build factories and create jobs, they create a multiplier effect and provide higher incomes.
   Level 2 (4–6 marks): More developed statements which explain the advantages and disadvantages of TNCs, e.g. TNCs create jobs for local populations who may previously have been unemployed. These employees usually receive higher wages than local businesses, which means that they have a higher income and a better standard of living, etc.
   Level 3 (7 marks): Three or more developed statements and **one named example** with at least **one** piece of place-specific detail, e.g. Nike is a TNC which makes sporting clothing and equipment. In 2010, 20 000 workers in Nike shoe factories went on strike for higher pay and in 2008 there were demonstrations in Indonesia when Nike decided to stop production in local factories and relocate, etc.

Geographical skills questions

1a Correct and accurate completion of population pyramid.
   [1 mark per correct bar.]

Total New Zealand population, 2006

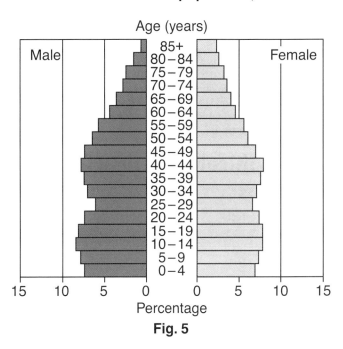

Fig. 5

Maori population, 2006

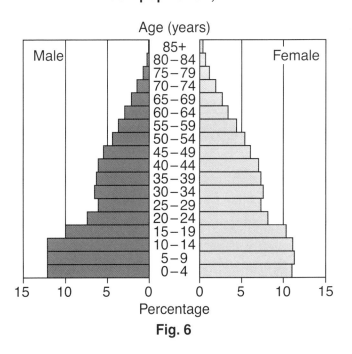

Fig. 6

Answers

1b i. greater **[1 mark]**

1b ii. less **[1 mark]**

1b iii. less **[1 mark]**

1c i. The proportion of Maoris in the New Zealand population is likely to increase **[1]**.

1c ii. The Maori population pyramid is youthful and shows a rapidly growing population **[1]** with a high birth rate and a relatively low death rate **[1]**.

2a. The relief in the area is steep **[1]** and made up of a number of undulating hills **[1]**. The land seems to be high **[1]** however, trees can be seen on the highest land and so is not so high that temperatures are too cold for trees to survive **[1]**. **[1 mark per point up to a maximum of 4 marks.]**

2b. On the highest land **[1]** and in valleys **[1]**.

2c. The steep slopes **[1]** which have been deforested **[1]** may encourage soil erosion.

Geographical investigations questions

1. Answer B is correct.

2. Answer A is correct.